T0351784

Sustainable Production

The Sustainability and the Environment series provides a comprehensive, independent, and critical evaluation of environmental and sustainability issues affecting Canada and the world today.

Anthony Scott, John Robinson, and David Cohen, eds., *Managing Natural Resources in British Columbia: Markets, Regulations, and Sustainable Development*

John B. Robinson, *Life in 2030: Exploring a Sustainable Future for Canada*

Ann Dale and John B. Robinson, eds., *Achieving Sustainable Development*

John T. Pierce and Ann Dale, eds., *Communities, Development, and Sustainability across Canada*

Robert F. Woollard and Aleck Ostry, eds., *Fatal Consumption: Rethinking Sustainable Development*

Ann Dale, *At the Edge: Sustainable Development in the 21st Century*

Mark Jaccard, John Nyboer, and Bryn Sadownik, *The Cost of Climate Policy*

Glen Filson, ed., *Intensive Agriculture and Sustainability: A Farming Systems Analysis*

Mike Carr, *Bioregionalism and Civil Society: Democratic Challenges to Corporate Globalism*

Ann Dale and Jenny Onyx, eds., *A Dynamic Balance: Social Capital and Sustainable Community Development*

Ray Côté, James Tansey, and Ann Dale, eds., *Linking Industry and Ecology: A Question of Design*

Edited by Glen Toner

Sustainable Production: Building Canadian Capacity

15 14 13 12 11 10 09 08 07 06   5 4 3 2 1

Printed in Canada on ancient-forest-free paper (100 percent post-consumer recycled) that is processed chlorine- and acid-free, with vegetable-based inks.

Library and Archives Canada Cataloguing in Publication

Sustainable production : building Canadian capacity / edited by Glen Toner.

(Sustainability and the environment)
Includes bibliographical references and index.
ISBN 13: 978-0-7748-1251-1
ISBN 10: 0-7748-1251-6

1. Industrial productivity – Environmental aspects – Canada. 2. Sustainable development – Canada. I. Toner, Glen, 1952- II. Series.

HC110.E5S92 2006   338.9′270971   C2006-901259-8

Canada

UBC Press gratefully acknowledges the financial support for our publishing program of the Government of Canada through the Book Publishing Industry Development Program (BPIDP), and of the Canada Council for the Arts, and the British Columbia Arts Council.

This book has been published with the help of a grant from the Canadian Federation for the Humanities and Social Sciences, through the Aid to Scholarly Publications Programme, using funds provided by the Social Sciences and Humanities Research Council of Canada.

UBC Press
The University of British Columbia
2029 West Mall
Vancouver, BC V6T 1Z2
604-822-5959 / Fax: 604-822-6083
www.ubcpress.ca

This book is dedicated to Professor Bruce Doern,
a mentor, an inspiration, and a friend to three generations
of Canadian public policy scholars.

At the heart of the assessment is a stark warning. Human activity is putting such strain on the natural functions of Earth that the ability of the planet's ecosystems to sustain future generations can no longer be taken for granted. The provision of food, fresh water, energy, and materials to a growing population has come at considerable cost to the complex systems of plants, animals, and biological processes that make the planet habitable ... Above all, protection of these assets can no longer be seen as an optional extra, to be considered once more pressing concerns such as wealth creation or national security have been dealt with ... Nearly two-thirds of the services provided by nature to humankind are found to be in decline worldwide. In effect, the benefits reaped from our engineering of the planet have been achieved by running down natural capital assets.

– Millennium Ecosystem Assessment,
Living beyond Our Means

Contents

Part 3: External and Internal Drivers of Sustainable Production

Conclusion

Acknowledgments

I would like to thank the authors for their work and patience over the lengthy gestation period of this volume. I would like to thank Randy Schmidt, our editor, and his colleagues at UBC Press, and the three anonymous referees who reviewed the manuscript and whose comments improved the volume.

I would also like to thank Elizabeth Dowdeswell, Alan Nymark, Bill Rees, Peter Harder, Peter Harrison, Aileen Carroll, Jerry Beausoleil, Robert Slater, Chris Henderson, Brian Emmett, Brian Guthrie, Richard Paton, Tony Hodge, Bob Page, Eli Turk, Dave Rodier, Stephen McClellan, Daniel Gagnier, Blake Smith, John Bannigan, David Runnalls, Arthur Hanson, Peter Behie, James Meadowcroft, Bea Olivastri, Brian and Dawn Davis, and Jim and Di McDonald for various contributions to this project during its development.

At Carleton University, I would like to thank Yixin Huang, Ryan Kuhn, Jeff Cottes, Bruce Doern, and the rest of my colleagues at CRUISE, the Carleton Research Unit in Innovation, Science and Environment.

The idea for this volume came from creating and teaching the Innovation, Science and Environment stream graduate seminar *Industrial Policy, Innovation and Sustainable Development* in the School of Public Policy and Administration at Carleton. The intent was to develop a Canadian source in an area in which the students were primarily reading literature written by American and European authors. Several generations of students in this class have read and critiqued earlier drafts of the chapters. Future classes will get the finished product.

Finally, I would like to thank my wife, Sylvia Haines, for her support and forbearance throughout the production of this book.

Introduction

If someone were to represent the Industrial Revolution as a retroactive design assignment, it might sound like this: Design a system of production that

- puts billions of pounds of toxic material into the air, water and soil every year
- measures prosperity by activity, not legacy
- requires thousands of complex regulations to keep people and natural systems from being poisoned too quickly
- produces materials so dangerous that they will require constant vigilance from future generations
- results in gigantic amounts of waste
- puts valuable materials in holes all over the planet, where they can never be retrieved
- erodes the diversity of biological species and cultural practices ...

The Next Industrial Revolution can be framed as the following assignment: Design an industrial system for the next century that

- introduces no hazardous materials into the air, water and soil
- measures prosperity by how much natural capital we can accrue in productive ways
- measures productivity by how many people are gainfully and meaningfully employed
- measures progress by how many buildings have no smokestacks or dangerous effluents
- does not require regulations whose purpose is to stop us from killing ourselves too quickly
- produces nothing that will require future generations to maintain vigilance
- celebrates the abundance of biological and cultural diversity and solar income.

– William McDonough and Michael Braungart,
"The Next Industrial Revolution," *Atlantic Monthly*

1
New Century Ideas and Sustainable Production

Glen Toner and David V.J. Bell

> In the end, sustainable development is not a fixed state of harmony, but rather a process of change in which the exploitation of resources, the direction of investments, the orientation of technological development, and institutional change are made consistent with future as well as present needs. We do not pretend that the process is easy or straightforward.
>
> – World Commission on Environment and Development, *Our Common Future*

> Can we move nations in the direction of sustainability? Such a move would be a modification of society comparable in scale to only two other changes: the Agricultural Revolution of the late Neolithic, and the Industrial Revolution of the past two centuries. These revolutions were gradual, spontaneous, and largely unconscious. This one will have to be a fully conscious operation, guided by the best foresight that science can provide. If we actually do it, the undertaking will be absolutely unique in humanity's stay on earth.
>
> – William Ruckelshaus, former head of the US Environmental Protection Agency

For thousands of years, human beings lived sustainably on this planet. This was not due to a carefully thought out philosophical commitment to the concept of sustainability.[1] Earlier societies lived more sustainably than we do out of sheer necessity, because nature imposed limits that the technology available to them could not easily expand or overcome. Societies that challenged these limits typically paid the high price of decline or, in some cases, extinction.[2]

The industrial revolution gave modern societies the illusion of omnipotence, the reassuring belief that the limits of nature could be transcended by technology. Increasingly this belief has collided with the disturbing reality of global climate change, ozone depletion, degradation of marine and terrestrial ecosystems, and the myriad other human-induced threats to the earth's life support systems. As the Millennium Ecosystem Assessment reminds us, it is increasingly apparent that for our species to survive into the future, humankind must learn to live differently on this planet.

If the challenge of making development sustainable is the overriding imperative of the twenty-first century, an integral aspect of this challenge is to shift our economy toward sustainable production. The purpose of this book is to strengthen the Canadian presence in the sustainable production debate while making a meaningful contribution to international discussions and scholarship on new approaches to industrial production.

Developments since *Our Common Future*

This debate was triggered by the publication in 1987 of *Our Common Future,* the report of the World Commission on Environment and Development (WCED 1987), which called for fundamental change to our systems of production and consumption (see in particular Chapters 3, 5, 7, 8, and 9). *Our Common Future* can be thought of as the "seed pod" from which the idea of sustainable development exploded onto the world stage.[3] The report emerged at a point in the late twentieth century when many people were growing increasingly anxious about the sustainability of both the existing industrial system and the global ecosystems that support life and commerce. The WCED concluded that "humanity has the ability to make development sustainable – to ensure that it meets the needs of the present without compromising the ability of future generations to meet their own needs" (WCED 1987, 8).[4] The winds of change carried the idea around the world in the period between 1987 and the 1992 United Nations Conference on Environment and Development (UNCED) in Rio de Janeiro (also known as the Earth Summit) and beyond. As Lafferty and Langhelle (1999, vii) have argued: "Though numerous critics have predicted a quick and definitive end to the idea, they have been proved decisively wrong. With the possible exception of democracy, there currently exists no more widely endorsed symbol of positive socio-economic and political change than 'sustainable development.' It is quite simply everywhere ... from the smallest local NGO, through all types of intermediate organizations and nation-states, to the United Nations, World Bank and European Union – it is what we are all (at least on paper) striving for."

The idea "seed" of sustainable development sprouted a wave of international actions beyond Agenda 21 and the UNCED Climate Change and Biodiversity Conventions, including the World Business Council for Sustainable

Development (WBCSD), the Johannesburg World Summit on Sustainable Development, and new research initiatives at the Organisation for Economic Co-operation and Development (OECD), the World Bank, and the International Organization for Standardization (ISO), among others. It also spurred a range of domestic responses in governments, companies, universities, and research organizations in various nations. How successfully the "seed" germinated and grew depended on how fertile the institutional soil was in these countries. A ten-country comparative study entitled *Implementing Sustainable Development: Strategies and Initiatives in High Consumption Societies* analyzed the post- Rio implementation effort and placed Canada at the top of the "cautiously supportive" category, the middle tier of engagement by the OECD countries studied (Lafferty and Meadowcroft 2000, 412; Toner 2000).

In Canada, therefore, the institutional soil was reasonably fertile for growing the seed of the sustainable development idea. The National Task Force on the Environment and the Economy was formed in October 1986 to respond to the WCED's May 1986 visit to Canada. Its 1987 report spawned the National Round Table on the Environment and the Economy (NRTEE) in 1989 (NTFEE 1987, 10). The Conservative government's 1990 Green Plan was the outcome of Prime Minister Brian Mulroney's commitment on the dais of the UN General Assembly to develop a sustainable development plan for Canada as recommended in *Our Common Future.*[5] The Liberal government of Jean Chrétien subsequently created the office of Commissioner of Environment and Sustainable Development and established the House of Commons Standing Committee on Environment and Sustainable Development. It also mandated the development, and updating every three years, of Sustainable Development Strategies by government departments. Canada is the first country in the world to legally require department-level sustainable development strategies. Along the way, the International Institute for Sustainable Development and Sustainable Development Technology Canada were created, and several new or amended pieces of legislation, such as the Canadian Environmental Protection Act and the Canadian Environmental Assessment Act, included sustainable development as a formal goal. Federal departments, such as Industry Canada and Natural Resources Canada, integrated sustainable development into their statutory mandates (for a more comprehensive analysis of this period, see Toner 2002 and Dale 2001).

Agenda 21, the forty-chapter manifesto adopted by all the world's nations at Rio, has been described as a "blueprint for the twenty-first century." It highlights the importance of sustainable production in Chapter 4, which links production to consumption, and points out that a major cause of the continued deterioration of the global environment is the pattern of unsustainable consumption and production, particularly in industrialized countries. A number of UN conferences since Rio have adopted the formulation of linking production to consumption, and have sought ways of making both more sustainable. From its inception, the UN Commission on Sustainable Development (CSD), which was established following the Rio Earth Summit to monitor progress in implementing Agenda 21, has identified "changing consumption and production patterns" as an "overriding issue" on its agenda.

Sustainable production can be thought of as a major subset of sustainable development. Its focus is on the systems by which industrial economies produce goods and services and the ways in which investment and production decisions are influenced by public policy. Sustainable production is dematerialized production. In its narrowest sense, it means minimizing energy and material extraction and throughput per unit of economic output. This entails, *inter alia,* product durability, energy efficiency, transforming wastes into raw materials, product redesign, and supplanting physical goods with knowledge goods. In its broader sense, sustainable production reduces material and energy use per unit of production while simultaneously improving environmental quality and improving social well-being. Sustainable production envisions an industrial system that would maximize resource efficiency, minimize environmental impacts, and replenish natural capital, while providing safe and satisfying employment opportunities. The public policy regime (tax system, regulations, information, incentives/subsidies) influences industrial practice and can be a facilitator of, or a barrier to, sustainable production by industry.

There are two aspects to the problem of production: what gets produced (sustainable products) and how it gets produced (sustainable production processes). These two aspects are not always recognized or appreciated. According to the Lowell University Center for Sustainable Production (2005):

> Sustainable Production is the creation of goods and services using processes and systems that are:
>
> - non-polluting
> - conserving of energy and natural resources
> - economically viable
> - safe and healthful for workers, communities, and consumers
> - socially and creatively rewarding for all working people.

Beyond the requirement that sustainable products be durable, easily recyclable, non-polluting, energy-efficient, and so on, further conditions must be met. The products should use appropriate packaging, have been subjected to a life cycle analysis, and be identifiable to potential consumers through a credible labelling procedure, such as a certification system developed by governments either on their own or in partnership with private and/or civil society organizations.[6]

As important as the kind of products produced in the new sustainable economy is the way in which these products are produced. One of the first industrialists to adopt sustainability as a core value for his company was Ray Anderson, founder and CEO of Interface Flooring. Anderson (1998, Chapter 5) describes sustainability as a mountain taller than Everest with at least seven "faces":

- waste elimination
- benign emissions
- closed loop production
- resource efficient transportation
- renewable energy
- redesigning commerce
- supportive ("sensitized") stakeholders.

Of these seven faces, four relate directly to the nature of industrial production processes, with emphasis on various ways in which production can be made cleaner and less material- and energy-intensive – in other words, more "eco-efficient."[7] While Robert Paehlke notes in Chapter 3 that sustainable production can be narrowly construed as "dematerialized production," he reminds us that a broader perspective is necessary to comprehend the many facets of sustainable production: "The achievement of broad sustainability involves sustaining our economy and the resources on which it depends, sustaining the quality of human life and therefore non-toxic air, water, and land, as well as a quality climate, and sustaining biodiverse wild nature (both for its own sake and to sustain both resources and the quality of human life). Every material input extracted from nature impacts one or more of these objectives."

To achieve the outcomes associated with this broader view will require more fundamental transformations of the economy than merely better production processes. This is where Ray Anderson's notion of "redesigning commerce" comes in. A sustainable economy will be more about services than products. Instead of owning products like carpets, customers will contract with manufacturers to provide them with the benefits of high-quality flooring by a lease arrangement that allows the manufacturer to assume control of and responsibility for the product "from cradle to cradle." As William McDonough and Michael Braungart (2002) argue, the goal of an "eco-effective" industrial system will be to design products and processes so that the non-toxic biological nutrients can be constantly recycled, whereas toxic technical "nutrients" can be constantly reused in closed-loop industrial processes. The aim is to ensure that the entire product is designed to be disassembled and recycled. This, in turn, will require different design features and a use of different materials.

But will customers accept these new arrangements? Interface Flooring's commercial "Evergreen" lease program has not been as popular as anticipated. Why not? What is needed to make it work? The importance of values and culture to the successful implementation of sustainable production is crucial. Anderson touches on this with his reference to "sensitized stakeholders." But who will make them sensitive to and supportive of these sustainability innovations? Who will help create an appropriate "culture of sustainability"? Pressure from large companies to "green the supply chain" is clearly one major driver (Willard 2002, 2005; Rowledge et al. 1999).

Once again, however, government must be factored into the sustainable production equation. In fact, government can and must play a number of different roles in helping "bend the curve" toward sustainability. It is important to ask how effectively Canadian governments are providing this kind of leadership. The public policy regime in Canada, and in other developed countries, is undergoing a period of transformation. While the existing taxation, regulation, information, procurement, and subsidization systems still, in many cases, support, or at least do not discourage, unsustainable production and consumption practices and habits, changes to these policy instruments are gradually appearing, partly in response to the growing critique of the policy system (Toner and Frey 2004). In February 2005, for the first time, a Canadian minister of finance, Ralph Goodale, appeared before the House of Commons Standing Committee on Environment and Sustainable Development to explain Finance Canada's position on using fiscal reform measures to achieve environmental and sustainable development goals. The movement toward ecological fiscal reform has been slow to gather strength in Canada (NRTEE 2002) but the Senate committee and several federal departments are now pressuring Finance Canada to employ economic instruments to help achieve Canada's Kyoto Protocol commitments and other sustainable development goals (Senate of Canada 2004, 2005). Finance Canada responded with a lengthy annex in the 2005 federal budget explaining the framework it will employ to evaluate environmental tax proposals (Canada 2005, 313-27).

But the question remains, when will human beings stop behaving like weekend visitors to the planet? What changes will be required in human behaviour, values, and systems to stop the pattern of unsustainable environmental, social, and economic development? The pursuit of a sustainable development trajectory requires the creation of sustainable production systems, which forces us to ask a series of tough questions.

How can –

1 industrial efficiency radically increase in order to extract four or even ten times as much productivity from the same throughput of energy and material resources?

2 product redesign reduce material content and eliminate or reduce toxic use during the product manufacturing, use, and disposal/recovery phases?
3 federal, provincial, and municipal policies support leading firms that are making investments and commitments and moving up the sustainable production ladder?
4 firms seize the opportunities offered by the triple-bottom-line orientation to ensure that they are internationally competitive and superior employers?
5 financial markets be encouraged to think beyond the next quarter and support innovative firms that are making investments that will improve society and the environment as well as the economy?
6 a tax system devised for an "empty world" where labour was scarce and resources were plentiful be reformed to serve a full world of 6-9 billion people where resources are increasingly scarce?

What –

1 values must change in order to overcome societal resistance to changes in our production and consumption habits?
2 are the barriers in the path toward a more sustainable future? Are they primarily technical, financial, policy, or psychological?
3 "disruptive technologies" now emerging will enable more economically efficient and environmentally restorative production and have an impact on society in the twenty-first century comparable to that of the automobile and personal computer in the twentieth century?

These are at least some of the key sustainable production questions being debated around the world, and Canadian public and private sector organizations, including universities, must strengthen their capacity to contribute to this debate. This book is a small contribution. While the critical literature on unsustainable production goes back to seminal works such as Rachel Carson's *Silent Spring* (1962) and the Club of Rome's *Limits to Growth* (Meadows et al. 1972), there has been a veritable explosion of both scholarly literature and industry/government/NGO publications on sustainable production following *Our Common Future* and the Rio Earth Summit. This literature has largely been generated in Europe and the United States, and much of it is cited in the following chapters. While there have been examples of Canadian firms and Canadian policy innovations discussed in this literature, they are very much the exception (see examples from Rowledge et al. 1999, and Hawken et al. 1999). A major advance in the development of a Canadian "literature" was the 2001 publication of a CD-ROM entitled *The Role of Eco-Efficiency: Global Challenges and Opportunities in the 21st Century* (Five Winds International 2001). The study sought to establish a body of Canadian experience through case studies of the innovative practices of fifteen companies, some

Canadian-owned, some foreign subsidiaries. Still, there is no question that there is a major lacuna in the Canadian literature on sustainable production.

This is not a traditional academic book in that only half the authors are university-based scholars. Interestingly, the fact that the others are practitioners underscores an important point. In Canada much of the cutting edge applied work is being done by consultants, who take ideas, concepts, and tools generated in other countries and apply them to the Canadian context on behalf of Canadian firms and government departments. Indeed, several of these authors are "change agents" who help companies develop the analytic tools and staff capacity to advance sustainable production within the firm and sector. Kevin Brady's work with the Product Sustainability Round Table connects him to a number of major international companies that are sharing experiences and learning collectively how to implement sustainable production practices and environmental management systems in business decision making. Blair Feltmate, Brian Schofield, and Ron Yachnin have helped create new capital market instruments for the Bank of Montreal while monitoring the changes in attitudes and practices of the capital markets for the Conference Board of Canada. At Stratos, John Moffet, Stephanie Meyer, and Julie Pezzack have done pioneering work in Canada in the area of corporate sustainability reporting. Working in partnership with federal and provincial departments and leading Canadian firms, they have produced a detailed benchmark survey of corporate environmental and sustainability reporting to gain a better understanding of current practices and to identify opportunities for improvement. The following section will highlight some of the central themes of the sustainable production debate identified by the authors of this volume.

Part 1: Sustainable Production and Its Context

Bob Masterson introduces many of the key ideas, concepts, issues, and arguments that characterize the current state of the debate. Masterson traces the debate back to *Our Common Future*'s formulation of sustainable development, which contained within it two key concepts, that of "needs" and that of "limits," "not absolute limits but limitations imposed by the present state of technology and social organizations on environmental resources and by the ability of the biosphere to absorb the effects of human activities" (WCED 1987, 8). In preparation for the Rio Earth Summit, the Business Council for Sustainable Development (it later changed its name to the World Business Council for Sustainable Development) created the concept of "eco-efficiency." Eco-efficiency sought to reduce inputs of energy, toxics, and materials into the production process and to minimize the impacts of product use on the environment. This concept was a breakthrough development for industry and, as Elkington has argued, functioned as a "Trojan horse" by getting environmental considerations into the boardroom. Nevertheless, it was criticized by others as being inadequate to meet the challenge of satisfying the needs of 9 billion people by the middle of the twenty-first century. Masterson explores the critique of eco-efficiency by McDonough and Braungart, Rees, and others.

Masterson provides a succinct and helpful summary of concepts such as industrial ecology, eco-effectiveness, biomimesis, the Natural Step, and ecological footprint, and the shift from a "goods production" to "service provision" economy, all of which call for radical resource productivity improvements. He then introduces the concept of corporate social responsibility (CSR). These concepts inform the analysis of other authors, with, for example, Moffet, Meyer, and Pezzack developing the theme of sustainability reporting in Chapter 7 and Feltmate, Schofield, and Yachnin focusing in Chapter 9 on changes in the capital markets to support sustainable production initiatives.

In Chapter 3, Robert Paehlke explores several sustainable production opportunities in a number of sectors, asking, "How do we gain sufficient foresight to intervene in terms of risk management without foreclosing economic and technological innovation?" He focuses on the choice and combination of policy instruments that improve risk decisions: "Innovation in sustainable production results from the intelligent anticipation of future trends and requires a climate that is likely to reward innovation." Because he is not convinced that the market alone will establish the necessary economic parameters to motivate and foster a sufficient level of innovation, he sees a crucial role for government in sustainable production, a role that in part includes getting resource prices right. Unlike pollution, which generally results from a limited number of particular processes or substances within particular industries, and which can be ameliorated with a limited number of innovations and interventions, sustainable production involves all aspects of the economy – hence, the scope of the policy challenges.

Paehlke constructs his sustainable production framework by critiquing the existing "common sense" model (which he calls "economism") that takes prosperity (as measured by GDP) to be the essential goal of society and the overwhelming objective of public policy. He borrows from Robinson and Tinker in arguing that two forms of decoupling are essential to move toward a sustainable society. One is the decoupling of economic output (GDP) from energy and material throughputs, especially the extraction of raw resources from nature. The other is the partial decoupling of social well-being from GDP per capita (that is, improving the quality of life faster than increases in wealth, or getting more for our money). The single bottom line orientation of economism is unduly one-dimensional, while sustainable development provides a framework that integrates increasing prosperity with increasing social well-being and enhanced environmental sustainability. Ultimately, industrial production and public policy must support the integrated goals of sustainable development and not the narrow goals of economism.

The information and communication technology (ICT) sector has a paradoxical impact in that it contributes to dematerialization via automation, telecommuting, and the substitution of physical goods with knowledge goods, while simultaneously generating problems with toxic material content and end-of-life management of outdated computer and other ICT equipment. (Keith Newton and John Besley, and David Wheeler, Kelly Thomson, and Michael Perkin, address other dimensions of this paradox in Chapters 4 and 5, respectively.) Paehlke is generally optimistic about the role that can be played by the ICT sector, along with other knowledge-based industries such as pharmaceuticals, health care, education, financial services, and the public sector, in a modern industrial economy because of their potential for growth without adding commensurately and proportionately to overall energy and material throughputs. He is less optimistic about the role of Canadian public policy here. He cites the OECD's 2000 economic survey of Canada to show that historically the leading sectors of the new economy, such as electronic equipment and industrial machinery, have been disadvantaged by Canadian public policy relative to the resource-based sectors, which have benefited from policies favouring resource development and use, particularly in the non-renewable resource sector (oil, gas, metals, and minerals).

Paehlke assesses recent sustainable production innovations in the transportation, construction, and urban planning sectors: infill housing in the urban core to slow suburban sprawl; the integration of material recovered from old tires into asphalt roads; the design of automobiles for disassembly and recycling. He also considers some of the market and policy barriers constraining the widespread adoption of these and other innovations. He also reviews a series of sustainability threats in the agriculture and forestry sectors in Canada, and some potential sustainable production opportunities for reversing many of these threats, including developments in the ethanol fuels area. Energy and resource pricing continue to be a major determinant of policy outcomes in his analysis. His hometown example of the multiple sustainable production benefits from a sophisticated piece of capital equipment (an industrial-scale laundry machine) built in Germany leads Paehlke to state: "One can only conclude that nations and industries that anticipate the resource limitations of tomorrow are likely to prosper in the face of global adversity ... Even if Canada has a relative abundance of water, trees, and energy it is in our best interest – environmentally, socially, and, in the long run, economically – to proceed as if we did not."

Part 2: The Knowledge-Based Economy, Social Capital, and Product Design

Keith Newton and John Besley argue provocatively in Chapter 4 that while the knowledge-based economy (KBE) has the potential to advance or impede sustainability, public policy innovations can help tilt the overall "social preference function" and "production possibilities frontier" in a positive direction. In their view, "certain features of the KBE – the emphasis on scientific and technological advance, continuous innovation, increased educational attainment, growing public concerns, the analytical and communication power of ICTs, dematerialization, and growing understanding of the mutually complementary objectives of eco-efficiency in firm performance – hold out the prospect of successfully addressing the admittedly daunting challenges."

They develop a model that argues that at the level of the firm, sustainable production should be a core component of the new business paradigm of the KBE. They explore the internal and external factors that motivate firms to adopt sustainable production practices and emphasize the growing salience of tools that focus on product life cycles, particularly on the crucial design stage. They note that many of the required models, tools, indicators, and resources (such as inventories of environmental technologies, best practices, case studies, etc.) reside in international institutions. The challenge of building Canadian capacity will depend, in part, on our ability to use, refine, and extend them for use in specifically Canadian situations (as was done by Stratos in *Stepping Forward* [2001] and *Building Confidence* [2003]).They explore the strengths and weaknesses of ICTs in promoting sustainable production, underscoring the positive role of ICTs in developing and disseminating knowledge, which helps build social capital and a supportive global constituency.

The role of the ICT sector in advancing sustainability and bridging the digital divide is also the focus for David Wheeler, Kelly Thomson, and Michael Perkin in Chapter 5. The digital divide has many dimensions (urban/rural; generational; economically engaged/disengaged, north/south). Wheeler and colleagues explore some of the positive and negative impacts of ICTs to date (and there are plenty of both), as background for assessing how business, government, and civil society organizations might partner to enhance social inclusion in both the developed and developing worlds. In fact, they argue, "because of the highly networked and relationship-dependent nature of the sector, there can be few industries of greater interest to social capital and sustainable development theorists than those reliant on ICTs."

They develop a conceptual framework that marries social capital and sustainability, and then apply the framework to some very interesting original research they have undertaken as part of the Sustainable Canada initiative. The research explores Canadian ICT sector attitudes toward sustainability and social capital in a number of Canadian-based ICT firms. One interesting finding from the research was that the internal drivers for pursuing sustainability (the need to deal with the waste generated by ICT hardware, and the values of ICT sector employees) were stronger than the external drivers (pressure from NGOs and governments). Interestingly, from a national competitiveness perspective, Canadian ICT opinion formers felt that there was greater leadership in Europe and that Canada needed stronger leadership from both government and the major sector players. Overall, they conclude that "there are no intrinsic barriers to ICT firms pursuing sustainability as a sector and/or as individual firms committed to creating economic, social, and ecological value for stakeholders. Of course, there are no guarantees that the sector in Canada will deliver such value, but at least there seem to be no philosophical barriers."

In Chapter 6, Carey Frey extends the focus of this section on technology and systems of innovation by analyzing the design professions and one of the dominant inventions of the industrial age, the automobile. Building on the eco-efficiency/eco-effectiveness debate introduced by Masterson, he develops an analytical framework for sustainable product design that highlights McDonough and Braungart's three broad sustainability categories: equity or social justice, economy or market viability, and ecology or environmental intelligence. "A new design must strive to optimize each of these three components of the sustainability triad ... The individual innovators who can combine the various criteria for sustainable production with their design and engineering skills will successfully drive this next industrial revolution." While innovations of this nature are possible, they are not yet the norm, and Frey critiques the dominant educational system for engineers and architects, which maintains the status quo, while underscoring the profound importance of the design stage on the life cycle impact of products and infrastructure.

Frey outlines the cultural, institutional, and economic factors that create barriers to sustainable production innovations in the training and employment of design professionals. He shares the viewpoint of Hawken and the Lovinses that no better high-leverage investment in the future could be made than "improving the quality of designers' 'mindware' – assets that, unlike physical ones, don't depreciate but, rather ripen with age and experience" (Hawken et al. 1999, 111). He highlights a recent example at the University of Michigan College of Engineering. The Concentrations in Environmental Sustainability, or "ConSenSus," program is designed to broaden the education of engineers by informing them about environmental regulations, policies, practices, and the implementation of clean technologies so that they can anticipate and help circumvent potential problems. A key pedagogical element of the ConSenSus program consists of case studies brought to the classroom by engineers from Ford, DaimlerChrysler, Dow Chemical, Pfizer, BP Amoco, and General Electric.

Ford's case study focused on the sustainable business case for rehabilitation of the firm's famous Rouge River manufacturing complex. William Clay Ford asked William McDonough to inspire and lead the remaking of the eighty-year-old "obsolete environmental catastrophe" into "the model of twenty-first century manufacturing." Based on McDonough's work in turning around the Rouge River complex, Ford asked McDonough to join the design team to develop Ford's Model U, which the company hopes will have the same impact on automotive production in the twenty-first century that the Model T had in the twentieth century. McDonough and Braungart wanted to use the opportunity to encourage the design team to shift from an eco-efficient strategy to pursuing eco-effectiveness, or positive environmental effects through intelligent design. This may seem like a tall order for the consumer product that has arguably had more impact on society and the environment than any other over the past century. Frey assesses the quite extraordinary design features of the Model U, including, for example, the supercharged hydrogen internal combustion engine, soy-based seating foams and tailgate resins, corn-based canvas roof and carpet mats, and many high-tech driver convenience, entertainment, and safety features. The development of the Model U as a concept vehicle raises a multitude of interesting questions that Frey explores – about eco-effective innovations and about the institutional, infrastructural, and attitudinal changes that will be required to support the full-blown diffusion of such innovations into the mainstream of society. Frey speculates that "the concepts of sustainable production, eco-effectiveness, and the Model U suggest that design professionals, a linchpin of the KBE, may be on the frontier of a transition from the ICT revolution at the end of the twentieth century to a new era of eco-innovation. There are, however, considerable technological obstacles that must be overcome before concepts such as the Model U are fully realized."

Part 3: External and Internal Drivers of Sustainable Production

The four chapters in this section focus on changes in the behaviour and mindsets of firms that are required for a sustainable future. Public policy is a prominent driver in creating the conditions for change in the norms and practices of firms. The key relationship is between industrial and environmental policy. The incentives, constraints, and signals provided by public policy and the emergence of a number of industry-generated tools and strategies are explored in these chapters.

In Chapter 7, John Moffett, Stephanie Meyer, and Julie Pezzack identify what they term the "twin dynamics" of sustainable production and consumption. These are the recognition by firms of: (1) the growing environmental pressures of an increasingly crowded and affluent world that are starting to threaten the ecological processes that sustain human life and underlie our economic prosperity; and (2) opportunities to improve corporate image and enhance market share, productivity, profitability, and competitiveness by adopting new, more environmentally benign modes of doing business.

Reflecting the debate outlined by Masterson, the authors align themselves with those who argue that an environmental protection/eco-efficiency approach is inadequate, in part because more efficient production may have the paradoxical effect of reducing the cost of products and thereby increasing levels of consumption. They are attracted to McDonough and Braungart's idea of "eco-effectiveness," which is a model of human industry that is "regenerative rather than depletive" and in which products work within "cradle-to-cradle" rather than "cradle-to-grave" life cycles. They also approvingly cite Hawken and colleagues' four dimensions of "natural capitalism" (1999): dramatically increasing the productivity of natural resources; shifting to biologically inspired production models; moving to a solutions-based business model; and reinvesting in natural capital. They identify the carpet company Interface as an example of a firm that has redesigned both its product and its production process to significantly reduce energy consumption, eliminate the use of almost all toxic substances, and reduce almost all waste from its factories. Interface also wants to shift from selling carpets to leasing "floor-covering services." This would enable it to maintain the carpet during its use phase and take it back at the end of its life. This provides the company with a powerful incentive to design products that are durable, easily maintained, and readily recyclable. The question addressed in Chapter 7 is: "What public policies are required to induce other companies to follow [Interface's] example, or to do even better?"

The authors argue that to become a widespread, underlying element of our business culture, sustainable production requires: (1) mutually coherent environmental and industrial policies focused on stimulating and disseminating innovation; (2) new mechanisms and measures for enhancing awareness of opportunities for change on both the demand and supply sides; and (3) price signals that eliminate inappropriate subsidies and start to incorporate environmental externalities. They explore four key policy measures that could help support these reforms: smart regulation; more use of "soft" instruments and partnerships; information disclosure programs; and ecological fiscal reform (EFR). In discussing EFR, they note that recent studies have estimated the value of the earth's ecosystem services at over $33 trillion a year, "yet, because the value of these services does not appear on any balance sheet, companies account at most for their use of resources, not for their impacts on ecosystem functions." EFR entails: (1) removal of all existing fiscal disincentives to environmentally sound practices; (2) use of eco-taxes to help internalize the true costs of production and consumption; (3)

selective use of incentives to encourage desired behaviours (e.g., accelerated capital cost allowances for energy and material-efficient technologies); and (4) development and use of new measures of progress that account more fully for environmental impacts than the current system of national accounts (NRTEE 2002, 2003).

One important new private sector sustainable production tool that the authors' consulting firm, Stratos, has helped pioneer in Canada is corporate sustainability reporting. The objective is to develop generally accepted sustainability accounting principles similar to principles and reporting frameworks that currently exist for financial statements and annual reports. Responding to the leadership of the Global Reporting Initiative to develop and disseminate guidelines for organizations to use for reporting on the economic, environmental, and social dimensions of their activities, products, and services, Stratos has taken a reporting methodology developed in the United Kingdom by SustainAbility and applied it to Canada. In *Stepping Forward: Corporate Sustainability Reporting in Canada* (2001) and *Building Confidence: Corporate Sustainability Reporting in Canada* (2003), Stratos has completed the first two in-depth benchmark surveys of corporate sustainability reporting in Canada. The third survey will be published in early 2006. This pioneering work, which resulted from a positive public/private partnership, has generated a better understanding of current reporting practices and their drivers and barriers, and has identified opportunities for improvement.

In Chapter 8, Mark Jaccard focuses on a new generation of regulatory instruments that can mobilize producers toward sustainable production. He is concerned about the wide gap that exists between our current environmental and industrial policies and the policies that are needed to shift us toward a more sustainable path. He arrays the traditional policy instruments (tools) along a continuum of "degrees of compulsoriness." His critique of the existing arrangement is that the most effective tools, such as command and control and environmental taxes, are the most compulsory and therefore least acceptable to industry, consumers, and politicians. The less compulsory information, moral leadership, and subsidy tools are highly acceptable to these constituencies, but because of their voluntary nature, they tend to be less effective. For Jaccard, the challenge is to develop policies that would achieve both effectiveness and acceptability to individuals and firms (i.e., not be seen as unfair or overly compulsory). He feels that EFR, levying of environmental taxes, and recycling of the revenue to reduce income and employment taxes and fund environmentally positive initiatives (such as vehicle feebates), may eventually emerge as a dominant policy instrument. He doubts, however, that current attitudes, particularly in North America, will allow this to happen across the broad policy spectrum in the short term.

Hence, to find a balance between accountability and effectiveness and to enhance economic efficiency, he proposes a focus on market-oriented, sector-specific regulations. He presents three examples of such tools and discusses their prospects for policy design and application: (1) sulphur dioxide emission cap with tradable permits in electricity generation, (2) the renewable portfolio standard (RPS) in electricity generation, and (3) the vehicle emission standard (VES). The success of the US sulphur dioxide cap and trade approach in reducing sulphur dioxide emissions and in stimulating additional health and economic benefits is outlined. The RPS has been adopted by half of the US states (DSIRE 2005) and several European countries and Australian states. Because polluting sources of electricity are currently not required to fully reflect the full environmental costs they cause, clean, renewable electricity technologies (solar, wind, running water hydro, wave and tidal generators, and geothermal energy) will require policy support until their costs fall as a result of commercialization and economies of scale. The RPS regulates the sellers of electricity, requiring them to procure a minimum percentage of the electricity they sell from renewable sources. There is no guaranteed electricity price for renewable electricity, only a guaranteed market share. Each producer decides whether to produce some renewable electricity themselves or purchase it from an independent producer. Trading mechanisms among producers mean those with the lowest costs will generate renewable electricity for the entire market. The market share targets vary with jurisdictions, and there are substantial penalties for non-compliance.[8]

The VES was established in 1990 by the California Air Resources Board. The policy requires that a minimum percentage of vehicle sales be of low-emission vehicles. Individual automakers must meet the standard as a fleet average of retail sales in California. A flexibility provision allows for trade between manufacturers so that the total vehicle fleet meets the standard even if individual manufacturers fall short. Automakers were given considerable lead times between target setting and target dates to provide enough time for the required technological change. Jaccard argues that the VES has played a key role in the emergence of revolutionary new vehicle technologies, notably electric/gasoline hybrids, battery electric, and fuel cell electric. Indeed, automakers are now trying to outcompete one another to capture this market, as reflected in recent research funding, commercialization efforts, and marketing strategies. A growing number of other US states have adopted the California standards and even enacted provisions to ensure that their standards adjust automatically if California modifies its VES.

After reflecting on the comparative advantage of such market-oriented, sector-specific regulations, Jaccard notes that, thus far, Canada's policy makers have been slow to adopt market-oriented regulations but that this may be changing. None of the "traditional" Canadian "barriers" of market size, jurisdictional coordination, and institutional inertia are insurmountable, and the federal and provincial governments are increasingly likely to experiment with what looks like an exciting new approach to balancing the diverse policy design objectives of cost-effectiveness and public acceptability.

Linked to the new tools identified by Moffett and his co-authors and Jaccard are new tools emerging in the capital markets that provide new insights into corporate behaviour and performance from a sustainable production perspective. In Chapter 9, Blair Feltmate, Brian Schofield, and Ron Yachnin discuss developments in the financial markets that are helping to drive advances in corporate sustainability reporting and new trends in investing. This work builds on their groundbreaking report for the Conference Board of Canada, *Sustainable Development, Value Creation and the Capital Markets* (Feltmate et al. 2001), which was the first major Canadian study to confirm that firms that have embraced sustainable development as a guiding principle are being recognized as superior investments. The adoption of sustainable production practices that minimize a firm's environmental impacts while simultaneously contributing to the economic and social development of the communities they operate in shows that these firms have a management orientation that goes "beyond compliance" and thinks long-term. Banks and insurance companies are acknowledging that firms managed in this fashion are a superior risk and hence their bank loan rates and insurance premium rates tend to be lower. Such firms usually practise the eco-efficiency credo of "doing more with less," seeking to reduce energy input, material requirements, and waste per unit of production. These companies tend to have better employee satisfaction and retention rates and easier access to new markets. All of these characteristics generally have a positive impact on share price appreciation.

It is not surprising then that the evidence is growing that funds and indices that invest in such firms are outperforming the market. The number of funds that employ a positive or inclusionary screen for SD practices is growing, and these funds are distinguishing themselves from traditional socially responsible investment funds that employ a negative screen for firms in the tobacco, gambling, alcohol, and weapons sectors. EcoValue 21™ in the US, the joint Swiss/US Dow Jones Sustainability Group Index, and the Norwegian/US Storebrand Scudder Environmental Value Fund are three of the best known and most successful. The Storebrand Scudder Environmental Value Fund employs nine eco-efficiency criteria to screen the environmental performance of a company: energy efficiency, global warming contribution, ozone depletion impact, material efficiency, product characteristics, quality of environmental management, toxic emissions, water use, and environmental liabilities.

Several changes in public policy are supporting this trend. Amendments to the UK pension fund regulations in July 2000 require funds to report whether they take account of the environmental, social, and ethical impacts of their investments. As a result, pension fund managers and financial analysts are seeking out companies whose shareholder value is enhanced, or at least protected, by prudent management of environmental and social risks. Feltmate and co-authors argue that perhaps the most far-reaching impact of these amendments is that they reverse the onus on pension fund managers by allowing sustainable development to be factored into the investment decision-making process without violating fiduciary responsibilities to maximize return: "In fact, recognizing the value creation and lower risk associated with sustainable development practices, it is now arguable that investment decisions made without assessing the environmental, economic, and social practices of companies may stand in violation of fiduciary responsibility."

While not as proactive as the UK initiative, regulators in the US, Ontario, and Manitoba made decisions in the 1990s that allow for the inclusion of SD criteria in mutual funds and pension plans. The UK changes have put pressure on firms trading on the London Stock Exchange to itemize and account for all their risks – financial, environmental, social, and ethical – and report on them at year-end in corporate reports. The Financial Times Stock Exchange (FTSE) announced the launch of the FTSE 4Good Index series to provide tradable benchmark indices. Other fund managers have launched funds based on the index. These are early days in the inclusion of sustainable development criteria in the orientation of the financial markets, and once again Canada is following developments started elsewhere. While a number of methodological issues need to be worked out to overcome remaining barriers to this process of moving SD investing into the mainstream of industry practice, this important new tool has the potential to significantly alter attitudes and behaviour in the corporate sector.[9]

Kevin Brady's chapter builds on the arguments presented in the preceding three chapters on the growing influence of external sustainable production drivers, by focusing on drivers and barriers internal to firms. Ultimately, for sustainable production values and changes to become embedded in the culture of a firm, the executive suite has to not only "buy in" but lead. As Brady notes, sustainable development has been characterized as a journey, not a destination. In the words of *Our Common Future,* "sustainable development is not a fixed state of harmony, but rather a process of change" (WCED 1987, 9). Indeed, Brady acknowledges that the sustainable production practice and value changes that flow from a firm's adoption of a sustainable development orientation may vary depending on whether the firm is in the resource extraction, financial, manufacturing, or retail sectors. One factor common to the journey along the sustainable development path, however, is this crucial role of senior management.

Brady analyzes the various dimensions of this large-scale change process, discussing the role played by crises, emerging analytical tools such as management systems and stakeholder engagement processes, "change agents" within the executive suite and board of directors, benchmarking, industry leaders outside the firm, and the increasingly sophisticated assessment tools for identifying both business risks and opportunities. Because he has both studied the corporate change process and worked closely with firms in a broad range of sectors attempting to embed sustainability practices and values into the organizational culture to ensure that real change takes place, Brady has a very strong sense of the barriers that exist and methods for overcoming them. Two are: (1) developing sustainable production terminology that the company leadership is comfortable with and that makes sense given the firm's size, location, and sector; and (2) overcoming the limitations of disciplinary training by finding ways to discuss sustainability that connects with the economists, accountants, lawyers, and scientists who occupy different functions within the executive suites.

In the concluding chapter, we integrate some of the key messages of the preceding chapters to show how the idea of sustainable production has grown over the past twenty years. We go on to explore further the role that public policy can and should play in supporting the growth of sustainable enterprise. We close by presenting four areas for further research.

Notes

1 Although it is true, of course, that most Aboriginal and so-called primitive societies have viewed nature very differently from "advanced" industrialized societies and typically treat nature with reverence and respect. In the Haudneshaunee language, for example, the word for "mother" means "you who gave me life." The concept of "Mother Nature" entails the notion of "you who makes all life possible."

2 There are many examples like that of Easter Island, often cited as a prototype of societies that ignore or attempt to transcend environmental limits and suffer the consequences.

3 Sustainable development was actually introduced into scientific discourse in the 1980 *World Conservation Strategy,* but remained largely within the domain of the scientific community until it was "popularized" by *Our Common Future.*

4 While many authors have attempted to develop modified definitions of sustainable development, none of these has gained the stature to supplant the WCED definition, which continues to be used by international organizations, government departments, companies, and civil society organizations. Sustainability is often used as a synonym for sustainable development. In Agenda 21, the terms "sustainability" and "sustainable development" were used interchangeably, and that is the case in this volume.

5 The Green Plan stated: "*Our Common Future* ... quickly captured the world's imagination with the concept of sustainable development [which it described] as activity in which the environment is fully incorporated into the economic decision-making process as a forethought, not an afterthought. It holds that resources must be treated on the basis of their future, as well as their present, value. That approach offers genuine hope of economic development without environmental decline" (Canada 1990, 4).

6 The German Blue Angel, the American ENERGY STAR, and the Canadian Eco-Logo certifications are examples.

7 Note the discussion by Bob Masterson in Chapter 2 of the "Factor 4" economy (possible with existing technology) and the much more ambitious "Factor 10" economy (which will require new technology and industrial and commerce redesign).

8 The government of Ontario introduced a similar tool when it committed to the development of new renewable sources of electricity generation with a goal of 5 percent by 2007

and 10 percent by 2010. In December 2004, the government announced ten new projects equal to 395 megawatts of renewable power; in March 2005, it announced a new request for proposals that will seek an additional 1,000 megawatts of renewable energy (Ontario Ministry of Energy 2005).

9 The National Round Table on the Environment and the Economy's Capital Markets and Sustainability project is advancing Canadian knowledge in this area. For example, studies done for this project explore corporate disclosure and capital markets, compare Canadian and UK pension fund transparency practices, and review the community investment sector in Canada (NRTEE 2005).

References

Anderson, R. 1998. *Mid-course correction: Toward a sustainable enterprise: The Interface model.* Atlanta: Peregrinzilla Press.

Boyd, D. 2004. *Sustainability within a generation: A new vision for Canada.* Vancouver: The David Suzuki Foundation.

Canada. 1990. *Canada's Green Plan for a healthy environment.* Ottawa: Minister of Supply and Services.

–. 2005. *The budget plan 2005.* http://www.fin.gc.ca.

Carson, R. 1962. *Silent spring.* New York: Houghton Mifflin.

Dale, A. 2001. *At the edge: Sustainable development in the 21st century.* Vancouver: UBC Press.

Database of State Incentives for Renewable Energy (DSIRE). 2005. *Rules, regulations and policies.* http://www.dsireusa.org/summarytables/reg1.cfm?&CurrentPageID=7.

Feltmate, B., B. Schofield, and R. Yachnin. 2001. *Sustainable development, value creation and the capital markets.* Ottawa: Conference Board of Canada.

Five Winds International. 2001. *The role of eco-efficiency: Global challenges and opportunities in the 21st century.* CD-ROM. Ottawa: Natural Resources Canada.

Hawken, P., A. Lovins, and H. Lovins. 1999. *Natural capitalism: Creating the next industrial revolution.* Boston: Little, Brown.

Lafferty, W., and O. Langhelle, eds. 1999. *Towards sustainable development: On the goals of development and the conditions of sustainability.* London: Macmillan.

Lafferty, W., and J. Meadowcroft, eds. 2000. *Implementing sustainable development: Strategies and initiatives in high consumption societies.* Oxford: Oxford University Press.

Lowell University Center for Sustainable Production. 2005. *What is sustainable production?* http://www.uml.edu/centers/LCSP/what.html.

McDonough, W., and M. Braungart. 1998. The next industrial revolution. *Atlantic Monthly* (October): 82-92.

–. 2002. *Cradle to cradle: Remaking the way we make things.* New York: North Point Press.

Meadows, D.H., D.L. Meadows, J. Randers, and W. Behrens. 1972. *The limits to growth: A report for the Club of Rome's project on the predicament of mankind.* Washington, DC: Potomac Associates.

Millennium Ecosystem Assessment. 2005. *Living beyond our means: Natural assets and human well-being.* http://www.millenniumassessment.org.

National Round Table on the Environment and the Economy (NRTEE). 2002. *Toward a Canadian agenda for ecological fiscal reform: First steps.* Ottawa: NRTEE.

–. 2003. *Environment and sustainable development indicators for Canada.* Ottawa: NRTEE.

–. 2005. *Capital markets and sustainability.* http://www.nrtee-trnee.ca/eng/index_e.htm.

National Task Force on the Environment and the Economy (NTFEE). 1987. *Report.* Submitted to the Canadian Council of Resource and Environment Ministers. September.

Ontario Ministry of Energy. 2005. *Renewable energy FAQs.* http://www.energy.gov.on.ca/index.cfm?fuseaction=renewable.faqs.

Rowledge, R., S. Barton, and K. Brady. 1999. *Mapping the journey: Case studies in developing and implementing sustainable development strategies.* Sheffield, UK: Greenleaf.

Senate of Canada. 2004. *The one-tonne challenge: Let's get on with it!* Report of the Standing Senate Committee on Energy, the Environment and Natural Resources. Ottawa. http://www.senate-senat.ca/EENR-EERN.asp.

–. 2005. *Sustainable development: It's time to walk the talk.* Report of the Standing Senate Committee on Energy, the Environment and Natural Resources. Ottawa: http://www.senate-senat.ca/EENR-EERN.asp.

Stratos Inc. 2001. *Stepping forward: Corporate sustainability reporting in Canada.* Ottawa: Stratos. Online: http://www.stratos-sts.com.

–. 2003. *Building confidence: Corporate sustainability reporting in Canada.* Ottawa: Stratos. Online: http://www.stratos-sts.com.

Toner, G. 2000. Canada: From early frontrunner to plodding anchorman. In *Implementing sustainable development: Strategies and initiatives in high consumption societies,* ed. W. Lafferty and J. Meadowcroft, 53-84. Oxford: Oxford University Press.

–. 2002. Contesting the green: Canadian environmental policy at the turn of the century. In *Environmental politics and policy in industrialized countries,* ed. U. Desai, 71-120. Cambridge, MA: MIT Press.

Toner, G., and C. Frey. 2004. Governance for sustainable development: Next stage institutional and policy innovation. In *How Ottawa spends 2004-2005: Mandate change in the Paul Martin era,* ed. G.B. Doern, 198-221. Montreal: McGill-Queen's University Press.

Willard, B. 2002. *The sustainability advantage: Seven business case benefits of a triple bottom line.* Gabriola Island, BC: New Society Publishers.

–. 2005. *The next sustainability wave: Building boardroom buy-in.* Gabriola Island, BC: New Society Publishers.

World Commission on Environment and Development (WCED). 1987. *Our common future.* New York: Oxford University Press.

Part 1
Sustainable Production and Its Context

2

From Eco-efficiency to Eco-effectiveness: Private Sector Practices for Sustainable Production

Bob Masterson

This chapter will review several major concepts that have influenced private sector sustainable production practices during the past two decades. The concept of sustainable development, as articulated by the 1987 World Commission on Environment and Development (WCED), is reviewed, along with the key implementation challenges identified during the 1992 United Nations Conference on Environment and Development (UNCED) in Rio de Janeiro (also known as the Earth Summit). The private sector's first response to those challenges, as reflected in the concept of eco-efficiency of the World Business Council for Sustainable Development (WBCSD), is presented along with a discussion of the range of critical reactions that accompanied it. This is followed by sections that discuss five more recent and far-reaching concepts that have been introduced by sustainable production theorists and leading private sector firms, including:

- a focus on radical resource productivity improvements
- the emergence of industrial ecology
- industry's shift toward being a supplier of service functions rather than products
- exploration of the wider, extra-environmental concept of the "sustainable corporation," which involves socioeconomic considerations along with the more traditional concerns of finance and the environment
- eco-effectiveness.

From *Our Common Future* to Rio

In its 1987 report, *Our Common Future,* the WCED defined sustainable development as "development that meets the needs of the present, without compromising the ability of future generations to meet their own needs" (1987, 43). For the WCED, sustainable development contained within it two key concepts, that of "needs" and that of "limitations," as imposed by the state of technology and social organization on the environment's ability to meet present and future needs. The key objectives noted by the WCED included: (1) extending economic growth that is more equitable and less material- and energy-intensive; (2) meeting essential needs for jobs, food, energy, water, and sanitation; (3) conserving and enhancing the natural resource base; (4) reorienting technology; and (5) merging ecological and economic considerations in decision making (1987, 49).

In the follow-up 1992 UNCED at Rio de Janeiro, two related aspects of human behaviour – the unsustainable patterns of production and consumption – were identified as the major causes of the ongoing deterioration of the global environment (UNCED 1992). Acknowledging its vanguard role in shaping the responses and direction of late-twentieth-century society, leading elements of the business sector reacted to the challenges and criticisms of UNCED through the articulation of a new concept, described as "eco-efficiency" (Schmidheiny 1992, 9).

In its initial configuration, eco-efficiency focused on seven key elements for business improvement, with the aim of reducing the demand for inputs into the production process and minimizing the impacts of production on the environment. In short, eco-efficiency was projected as an internally driven concept focusing on issues of corporate economic efficiency with positive environmental co-benefits (UNEP and WBCSD 1997, 3).

While the concept of sustainable development has achieved normative status in public and business discourse since 1987 (Lafferty and Langhelle 1999, 26; Lafferty and Meadowcroft 2000), the overarching desire for material development and economic growth, sustainable or otherwise, remains dominant. In recognizing the reality of a burgeoning population with a ravenous appetite for energy and material inputs to meet its needs, criticism of the eco-efficiency concept was widespread. This criticism noted that while the evolutionary response inherent in the eco-efficiency concept was sufficient from the internal and shorter-term corporate point of view, it was insufficient from an external and longer-term societal vantage point, and thus incapable of addressing the central challenges relating to sustainable development. According to Rees (1995), these challenges can be summarized as the requirement to accommodate both rising material expectations, and a 50 percent increase in population over the next fifty years, while simultaneously reducing total throughput.

The relative ease with which the central challenge of sustainable development can be articulated belies the enormity of the efforts that will be required to meet the challenge, given that by 2050 a world with 9 billion people will likely already be testing natural limits in one, if not several, of the following sectors: fresh water, forests, rangelands, oceanic fisheries, biological diversity, and global atmosphere (Brown and Flavin 1999, 11; UNEP 2005).

In considering these challenges, leading private sector firms have begun to move beyond the eco-efficiency concept as originally conceived in response to the UNCED challenge. Among the more interesting and meaningful responses initiated have been:

- a move toward radical (90 percent) resource productivity improvements
- industrial ecology
- industry's shift toward being a supplier of service functions rather than products
- exploration of the concept of "corporate social responsibility," which involves socioeconomic considerations along with the more traditional concerns of finance and the environment
- consideration of a broader eco-effectiveness ethos in industrial design.

Radical Resource Productivity Improvements

While some observers suggested that eco-efficiency represented a useful transitional strategy and has functioned as sustainability's Trojan horse by getting environmental considerations into the boardroom (Elkington 1998, 7), the initial configuration was criticized by many observers. These critics argued that the efficiency improvements at the heart of the eco-efficiency concept would be insufficient to counter the absolute impacts of the ongoing rapid growth in population and the corresponding energy and resource demands.

By some estimates, humanity's harvest of nature is already approaching half the available output at a time when economic activity is expected to increase by a factor ranging from 5 to 10 in the next fifty years[1] (Rees 1995; Hart 1997; Brown and Flavin 1999; UNEP 2005). The applied technology approach of eco-efficiency was criticized as also contributing to increased net consumption by putting economy first and making the extraction and use of resources ever more productive (see Chapter 7 for more on this point).

As economist Herman Daly argues, productivity improvements more generally have created a unique juncture in human history where the limits to increased prosperity are not the lack of financial or man-made capital but rather the lack of natural capital. The limits to increased fish harvests are not inefficient boats but rather the availability of fish due to overefficient boats, while the limits to pulp and lumber production are not inefficient harvesting and processing machinery but rather the continued availability of plentiful forests (Daly 1994). Following a similar theme, Lester Brown has been particularly effective in articulating the case that human demands are outstripping the earth's natural capacities, especially in areas of food production (Brown and Flavin 1999; Brown 2005).

McDonough and Braungart (1998), in a widely reproduced article originally published in *Atlantic Monthly,* were especially critical of the eco-efficiency concept, declaring that "eco-efficiency ... is not a strategy for success over the long term. [It] will not save the environment ... [it] will in fact achieve the opposite – it will let industry finish off everything quietly, persistently, and completely."

One of the key criticisms of the earlier configuration of eco-efficiency was over the lack of quantifiable targets in the face of well-quantified challenges. The early eco-efficiency targets of "reduce," "enhance," or "increase" (UNEP and WBCSD 1997, 3) were not considered adequate and were assessed as being unable to address the fundamental cultural values or the evident growth ethic that are leading to crisis (Rees 1995).

In considering the shortcomings of the eco-efficiency concept, some analysts began articulating more far-reaching assessments of what is possible and what is ultimately required. In their groundbreaking work, von Weizsäcker and colleagues (1998) showed that Factor 4 improvements to the economy could be achieved, at net economic benefit, through a doubling of wealth and a halving of resource and energy use. Others concluded, however, that given the growth of the world's population over the next century, and industrialized nations' per capita consumption rates of at least five times those of developing nations, reducing materials use by 50 percent globally implied a reduction of material intensities by at least a factor of 10, or 90 percent, in OECD nations (Rees 1995; von Weizsäcker et al. 1998, 244).

The WBCSD did not take criticism of its invented concept lightly, and suggested that McDonough and Braungart, as well as other critics, "do not understand (eco-efficiency)." Nevertheless, the WBCSD responded to the multitude of criticisms by substantially revising its approach to resource productivity. This is evidenced in its latest incarnation of the eco-efficiency concept, now described as being "achieved by the delivery of competitively priced goods and services that satisfy human needs and bring quality of life, while progressively reducing ecological impacts and resource intensity throughout the life cycle, to a level at least in line with the earth's estimated carrying capacity" (WBCSD 2000).

In rethinking the concept, leading proponents have identified four aspects that distinguish the current incarnation from the earlier versions. Rather than focusing on "reduce," "enhance," and "increase," eco-efficiency is now based on the following (Holliday et al. 2002, 83-84):

- "dematerialization" – substituting knowledge flows for material flows
- "production loop closure" – design and operation of zero-waste facilities, wherein every output is returned as a nutrient to natural systems, or becomes an input in the manufacture of another product
- "service extension" – moving from a supply-driven to a demand-driven economy and taking responsibility for the recovery and reuse of all product materials when they are no longer needed or wanted by the customer

- "functional extension" – manufacturing smarter products and selling services to enhance the products' functional value.

Industrial Ecology

In recent years, those concerned with the challenges of achieving sustainable development have also begun to look to nature itself for inspiration and guidance on "dematerializing" and "detoxifying" the industrial economy. The resulting process, involving an examination of the flow, use, and transformation of material and energy resources in industrial and consumer activities, and their corresponding effects on the environment, is referred to as "industrial ecology" (White 1994). As a guiding framework, industrial ecology implies three distinct principles (Lowe 1996, 445):

- All industrial operations are natural systems that must function as such within the constraints of their local ecosystem and the global biosphere.
- The dynamics and principles of ecosystems offer a powerful source of guidance in the design and management of industrial principles.
- The ultimate source of economic value is the long-term viability of the planet and its ecosystems.

McDonough and Braungart (1998) have summarized and popularized these guiding principles of industrial ecology as "waste equals food," "respect biological diversity," and "use solar energy."

Natural systems are admired for their beauty and seeming simplicity. Yet, as any biologist can attest, individual life forms, no matter how simple, are dependent on a much wider, more complex ecosystem for their survival. The objective of industrial ecology, then, is to begin moving from the present linear production system of "resource to product to waste" to a closed-loop model more closely resembling the cyclical flows of energy and resources within ecosystems (Lowe 1996, 438).

All creation processes involve the use of materials and energy that do not remain embedded in the final output. In natural approaches to creation, however, such materials become feedstock for other creation processes, as eventually the final output itself does. The "zero-emissions" concept attempts to capture the essence of this effective natural process and apply it to industrial production and consumption practices. In learning from nature, industrial practitioners of the "biomimesis" concept study these basic principles of natural systems and emerge with applications to meet the needs of mankind (Lowe 1996, 459). At the individual product level, biomimesis may lead to super-strong materials patterned on spider's webs, or manufactured enzymes that mimic those of the rainforest and effectively snap the strong carbon bonds in plant cells, releasing their sugars and forming valuable "biofuels."

In reflecting on her own pioneering work in this area, Janine Benyus (1997) suggests: "We don't need to invent a sustainable world – that's already been done! It's all around us in nature. We only need to learn from its success in sustaining the maximum wealth and the minimum materials flows."

At the facility level, the natural concept of "by-product synergy" is employed so that the waste heat from industrial power plants can be used to heat homes rather than nearby bodies of water. At a much larger level, new-century industrial parks such as that in Kalundborg, Denmark, are practising "industrial symbiosis" as five core industries – a power station, a refinery, a Gyproc factory, a pharmaceutical company, and the municipal water and heating company – interact so that the residuals of one become feedstock for another (Lowe 1996, 458-61). While the Kalundborg site evolved as a brilliant contrast to typical industrial production practices, it should be noted that "design for the environment"' may be a more effective approach to industrial and product design, as 80-90 percent of the cost of a typical product, and 80 percent of its environmental impact, is determined during the design stage (Elkington 1998, 361).

In "design for the environment," the aim is to systematically consider environmental, health, and safety objectives over the full product and process life cycle up front during the design process (Fiskel 1996; Charter and Tischner 2001; Chapter 6). Key focus areas include incorporating the zero-waste concept inherent in industrial ecology, and reducing energy needs so that renewable energy sources become more practical for operational use. As McDonough and Braungart suggest, in order to reach zero-emissions states of industrial production, "you don't filter water or exhaust gases, instead, you put the filter in your head and design the problems out of existence." Further, and in relation to public policy, their assessment of the application of industrial ecology and design for the environment means that "in the next industrial revolution, regulations will be seen as a signal of design failure" (McDonough and Braungart 1998).

On a more modest level, several examples exist in which Canadian firms are working to close the production loop, turning customers into suppliers. Alcan and Ford, in the first partnership of its kind in the North American automobile sector, have established a recycling plan through which recovered scrap aluminum from Ford operations is returned to Alcan for recycling directly back into auto body sheet metal. In another first, Noranda extracts valuable metals from Hewlett-Packard computers and other information and communication technology (ICT) products at the end of their working lives and resells the metals. Weyerhaeuser and others in the forest products sector are also establishing customer partnerships that close the loop between the manufacturer and consumer. Over the past five years, Stora Enso, another major operator in the forest products sector, has reduced the amount of waste to landfill by over 50 percent and has established a longer-term goal of having 100 percent of high-volume wastes beneficially reused (Stora Enso 2003).

While industrial ecology can lead to fuller use of raw materials and a shift toward renewable energy sources, several barriers prevent widespread implementation of the ideas and devices in these toolkits. Competitive pressures can lead to pushing ahead with new products and practices without adequate consideration of the precautionary principle, which argues for erring on the side of human and environmental safety in the absence of scientific clarity. Others argue that a limiting element for sustainable industrial production is the absence of measurement standards or "sustainability metrics" that can act as a guide (National Academy of Engineering 1999, 195). Consistently, though, analysts argue that "in commodity after commodity critical to the long term viability of nature, market prices are a fraction of their true costs" (Durning 1998, 244). Thus, private sector progress on sustainable development remains inextricably linked to broader public policy.

In addressing both the precautionary principle and the lack of sustainability metrics, Karl-Henrik Robèrt's four "Natural Step" principles act as an industrial ecology compass to guide the questioning firm toward sustainability. These Natural Step principles (Robèrt 2002) argue that:

- substances from the earth's crust must not be extracted at a rate faster than they redeposit into the earth's crust
- substances must not be produced by society faster than they can be broken down in nature or deposited in the earth's crust
- the physical basis for nature's productivity and diversity must not be allowed to deteriorate
- there must be a fair and efficient use of energy and other resources to meet human needs.

While eloquently stated, such principles may seem overly simplistic and of little use. Yet many corporations, including major names such as Home Depot, BP, Interface, Nike, and BHP Billiton, have begun the process of incorporating these same principles into their own operating codes and practices.

At the macro, national level, industrial ecology indicators for sustainability also exist. One such indicator is the "ecological footprint." Created by Wackernagel and Rees (1996), the ecological footprint attempts to evaluate societal activities in relation to their local ecosystem carrying capacity, by determining the area of productive land and water required to produce, on a continuous basis, all the resources consumed and to assimilate all the wastes generated by a given population. This concept is useful in showing just how far from sustainability our national systems are by suggesting that six planet Earths would be required to provide the world's population with the same production and assimilative capacities enjoyed by the citizens of

Canada. More recently, Canada's National Round Table on the Environment and the Economy (NRTEE) has sought to expand the existing series of national economic accounts to include measurements related to natural and human capital (NRTEE 2003). The initial set of indicators chosen by NRTEE will assist in understanding aggregate performance in moving Canada toward more sustainable development.

Unfortunately, however, industrial ecology does not contain within itself a method for correcting the prices of natural resources and industrial wastes, which are currently blind to most ecological and social costs. Finding a way to correct resource and waste prices and "make them tell the truth" (Hawken 1993) will be necessary, however, for fuller implementation of industrial ecology as well as for facilitating a shift by industry from production of goods to provision of functions.

The Shift from "Goods Production" to "Service Production"

Closing the materials loop through by-product synergy and other industrial ecology approaches is a necessary, but not a sufficient, condition to slow the rapid and unsustainable flows of materials and energy through economies. Also required is a radical shift from an economy where industry and consumers see wealth as a function of the sale and acquisition of goods, to an economy where the continuous provision and receipt of quality, utility, and performance delivers well-being (Hawken et al. 1999, 10). Through such a "redesign of the business model," profitability can be decoupled from the throughput of goods, as the basic business goal will involve selling results rather than equipment (Hawken et al. 1999). Leading firms have pioneered this transition from the sale of goods to the provision of results and have demonstrated its profitability and results.

Envisioning the capacity of this "new-century transition" to achieve the sustainability objectives outlined by the WCED is dependent on the acceptance of two basic but nonetheless jarring realizations. The first is that companies are not in business for the purposes of selling fans, motors, and condensers, nor are companies specifically in business for the purpose of contaminating soil, water, and air. Instead, companies are in business for the primary purpose of profit making. Second is the realization that customers generally do not make purchases because of what a product *is;* they make purchases because of what a product *does*. These realizations lead to the conclusion that excessive consumption of energy- and resource-intensive, but labour-lean, products is not required to deliver the services people want. The challenge and opportunity for business in the new century is to maximize the provision of such services while minimizing the production of goods (Roodman 1999, 169).

This shift from the production of goods to the provision of services begins with industrial ecology's "Design for Environment" techniques of incorporating materials and designs that are durable on the one hand and modular on the other. Products that are both durable and modular are then easier to recondition and remanufacture.

In the current economy, where products become obsolete shortly after purchase as a result of fast-moving advances in technology, products designed for the environment are also easily upgradable. Reconditioning, remanufacturing, and upgrading all realize the customer's true and basic needs. They also provide opportunities for long-term success and competitive advantage for firms that incorporate such practices, since even components that survive through just three recyclings achieve, on average, 60 percent lower per-unit cost reductions (Lowe 1996, 484). As well, in today's competitive marketplace, business is no longer able to count on consumers' brand loyalty for future sales. Even those serving larger commercial and institutional customers report losing more than half their clients in five years time (Elkington 1998, 231). By initiating a switch from strictly a sales operation to a leasing operation, however, business can build long-term customer relationships, and therefore success, through the continuous maintenance and improvement of highly durable products, which the company itself maintains ownership and end-of-life responsibility for. Interface Inc. has been one of the early pioneers in this ongoing transition away from the production of goods to the provision of services. With its "Evergreen" leasing arrangement, Interface has established what it calls "the first perpetual lease for carpet" (Anderson 1998). Through such a leasing and remanufacturing process, Interface has begun to ask the fundamental question necessary for sustainability: "What business are we in anyway?" For Interface this has meant a shift from being a seller of carpets to being a provider of flooring services. The leasing of flooring services leads not only to improved customer relationships but also to an improved bottom line. With each remanufacturing of their own "perpetual assets," Interface improves its competitive position relative to "once through" manufacturers, whose frames of reference measure success by the speed at which they can lose control of products and materials (Frankel 1998, 175).

Other major corporations with similar standing in their own sectors have also begun to ask the same fundamental question about which business they are truly in. Carrier, once the largest seller of air conditioning equipment, is now the world's largest supplier of "coolth (i.e., cooling) services" (Hawken et al. 1999, 135). The concept is catching on, partially driven by external challenges from a more critical and knowledgeable public. As a result, companies are finding that their responsibilities for a product cannot be shrugged off at the factory gate or point of sale. Instead, they are being encouraged, and increasingly forced, to take cradle-to-grave responsibility for the environmental and social impacts of their activities as well as those of their suppliers and customers (Elkington 1998, 9, 189).

Rapidly advancing communications technologies are also driving the transition, as can be seen by Canada Post's launching of its "epost" service. In an era of increasing competition and personal electronic communications, Canada Post has recognized that its future is not in delivering bills and letters door to door but in providing a secure and reliable portal through which parties can confidentially exchange information.

Xerox, once the single largest seller of photoduplicating machines, has made the transition and is calling itself "The Digital Document Company." Much like Interface, Xerox uses its leased assets (copiers) as a source of high-quality, low-cost parts and components for new machines. It has developed a well-run infrastructure for taking back not just leased copying machines but also the normally consumable and discardable components, such as toner cartridges, and, equally important, the component packaging (Xerox 2004).

The transition toward providing the services of products rather than the products themselves can also assist in addressing unemployment, a problem searching for a solution nearly the entire world over. Even the simple reconditioning of an automobile to make it last ten years longer requires 42 percent less energy and 56 percent more labour than manufacturing a new car (Rees 1995). A more robust approach, involving the leasing of product services, leads directly to a decentralized and skill-based service economy where economic value is based on utilization of products rather than exchange. As the concept expands, decentralized, labour-intensive service centres would create the needed jobs for those employees no longer needed in the centralized and highly mechanized industries involved in the extraction and energy-intensive transformation and production of finite resources (Lowe 1996, 455). Opportunities such as this, which lead to not only "dematerialized" and "'detoxified" but also high-employment economies, deserve serious attention in a world in which nearly 10,000 new people are born each hour.

Corporate Social Responsibility

As noted by Holliday et al. (2002), in its early manifestations sustainable development was largely an environmental agenda. More recently, stakeholders have been drawing attention to the social side of the concept and are increasingly challenging companies on a wider range of issues, including corporate governance and business ethics, human and labour rights, material liabilities, community investment, and relations with indigenous peoples, among others.[2] Moreover, interest in these diverse issues has been accompanied by overarching demands for increased openness, transparency, and accountability, which have translated into actionable calls for fuller and more effective stakeholder involvement in firms' decision-making processes, and for firms to take responsibility for and publicly report on their socioeconomic performance in addition to their financial and environmental performance.

A growing number of companies are responding to these challenges and demands through increased disclosure, including reporting on corporate social responsibility issues, and through the development of more comprehensive strategies and management systems.[3] This is reflected in the rapid growth in corporate "sustainability" reporting in recent years. In reflecting on the results of its 2005 update, the management consulting firm KPMG suggests that "corporate responsibility reporting in industrialized countries has clearly entered the mainstream," with over 50 percent of the Global Fortune 500 now publishing such reports (KPMG 2005).

Canadian firms have also begun to take up the reporting challenge. According to a survey by Stratos Inc. (2005), 25 percent (55 of 220) companies listed in the Toronto Stock Exchange Composite Index (TSX) publicly reported, in a substantial manner, on environment, health and safety, community, social responsibility, or sustainability issues in 2004. Internationally, this places Canada third, behind only Japan and the United Kingdom in terms of the percentage of top companies (by revenue) producing sustainability reports (KPMG 2005).

With the increase in the number of firms reporting, firms and stakeholders have identified the need for a more standardized approach. As a result, the Global Reporting Initiative (GRI) has emerged as the leading framework for voluntary sustainability reporting. A multi-stakeholder initiative, the goal of GRI is to raise sustainability reporting to the level of rigour and comparability of financial reporting (GRI 2005).

Within their reports, leading corporations are also beginning to address investor and other stakeholder concerns over firms' material risks and other long-term liabilities. Driven by stakeholder, shareholder, and investor actions, companies are now coming clean on their long-term liabilities associated with issues such as contaminated sites, greenhouse gas emissions, land restoration, hazardous substance (e.g., asbestos) inventories, and shortfalls in corporate pensions (see Chapter 9). Firms are also moving well beyond merely reporting on their current performance and have begun to realign their internal management structures to ensure continuous improvement toward established sustainability goals. Internal realignment of management systems has also begun to focus on the social component of sustainable development (see Chapter 10).

Unsurprisingly perhaps, given its historical record and its need to maintain a social licence to operate, the mining sector, both in Canada and internationally, has been at the forefront of efforts to integrate social and environmental dimensions into its internal management process. Global mining giant BHP Billiton has implemented a comprehensive Health, Safety,

Environment and Community Management System at all of its facilities. The system builds on a set of fifteen management standards, covering issues such as product stewardship, business conduct, human rights and indigenous affairs, and crisis and emergency management. Facilities' performance against these standards is monitored, audited, and reviewed to identify trends, measure progress, and drive continuous improvement, and the results are publicly reported on an annual basis (BHP Billiton 2002). The Mining Association of Canada has embarked on its own "Towards Sustainable Mining" initiative, which is being driven by fourteen similarly comprehensive principles (Mining Association of Canada 2005).

Eco-effectiveness

Through their work with leading firms, such as Nike and Ford, McDonough and Braungart have refined the ideas they first articulated in the 1998 *Atlantic Monthly* article, and published them in *Cradle to Cradle: Remaking the Way We Make Things* (2002). Here they propose a five-step process that encompasses all of the above concepts and more.

From Step 1 to Step 3, McDonough and Braungart challenge designers to progressively weed out and replace existing products' ecologically harmful ingredients and move from redefining products as "free of" one particularly dangerous substance to selecting all product ingredients from a well-defined menu of safe materials. At Step 4, designers are tasked with creating products and systems that are not simply "less bad" but that actually generate a broad spectrum of positive effects, including the creation of ecological, social, and economic value. At Step 5, designers are asked to consider more esoteric questions: "How might my product fulfil people's wants, needs and loves? Are my current business practices the best way to provide my service to customers? What service am I providing, anyway?" (McDonough and Braungart 2002).

Rather than striving for eco-efficiency, doing more with less environmental impact, eco-effectiveness seeks to produce "the right thing, the right service, the right product, the right system – rather than making the wrong thing less bad." As such, the concept of eco-effectiveness challenges the twenty-first-century industrial designer to move beyond considerations of sustainable production and to begin intentionally and meaningfully treating the challenges posed by the WCED with the seriousness they require.

Conclusion

As Roome (1998) notes, only a select few companies currently operate at a level of excellence implied by adoption of the concepts and practices discussed in this chapter. Nevertheless, while the impacts of these changes may be limited to date, there is no question that a trend is underway and that firms are increasingly working to minimize risk and increase business value by incorporating these considerations into their business management systems. In his seminal 1998 work, *Cannibals with Forks,* Elkington foresaw these trends and argued that, to ensure that corporate governance structures remained relevant in the twenty-first century, they would need to focus on new questions such as "What business are we in? Who should have a say in how the corporation is run? What changes are needed so that there is not a division between profits and sustainability?" (Elkington 1998). Such questions are proving to be not merely rhetorical, and are in fact already guiding the management and practices of leading firms.

Notes

1 Rees quotes a 1986 study by Vitousek and colleagues, which concluded that human beings directly or indirectly appropriate almost 40 percent of net terrestrial photosynthesis to their own use. Rees argues that "such data show that humanity's harvest of nature is already approaching half of the available output at a time when the prevailing international development model anticipates a five-to-tenfold increase in world economic activity by the time the human population stabilizes (at about ten billion) toward 2050" (see Rees 1995).

2 In a 2001 poll conducted by Environics, over half of the 1,000 respondents agreed that a company's social and environmental performance was as important as its financial performance. In a separate poll, conducted for Canadian Democracy and Corporate Accountability Commission (2001), 72 percent of respondents agreed that executives have a responsibility to take account of the effect of business decisions on employees, communities, and country.

3 In a 2002 PricewaterhouseCoopers survey of CEOs of leading US companies, 89 percent of CEOs polled identified sustainability as a growing business issue.

References

Anderson, R. 1998. *Mid-course correction: Toward a sustainable enterprise: The Interface model.* Atlanta: Peregrinzilla Press.

Benyus, J. 1997. *Biomimicry: Innovation inspired by nature.* New York: William Morrow.

BHP Billiton. 2002. *Annual health, safety, environment and community report.* Melbourne: BHP Billiton.

Brown, L. 2005. *Outgrowing the Earth: The food security challenge in an age of falling water tables and rising temperatures.* Washington, DC: Earth Policy Institute.

Brown, L., and C. Flavin. 1999. A new economy for a new century. In *State of the World,* ed. L. Brown, C. Flavin, and H. French, 3-21. New York: Norton.

Canadian Democracy and Corporate Accountability Commission. 2001. *The new balance sheet: Corporate profits and responsibility in the 21st century.* Toronto: CDCAC.

Charter, M., and U. Tischner, eds. 2001. *Sustainable solutions: Developing products and services for the future.* Sheffield, UK: Greenleaf.

Daly, H. 1994. Operationalizing sustainable development by investing in natural capital. In *Investing in natural capital,* ed. A. Jansson, ch. 2. Washington, DC: Island Press.

Durning, A. 1998. Changing world view. In *The World Watch reader on global environmental issues,* ed. L. Brown and E. Ayres, ch. 5. New York: Norton.

Elkington, J. 1998. *Cannibals with forks: The triple bottom line of 21st century business.* Gabriola Island, BC: New Society Publishers.

Environics. 2001. *Corporate social responsibility monitor 2001.* Toronto: Environics.

Fiskel, J., ed. 1996. *Design for the environment: Creating eco-efficient products and processes.* New York: McGraw-Hill.

Frankel, C. 1998. *In Earth's company: Business, environment and the challenge of sustainability.* Gabriola Island, BC: New Society Publishers.

Global Reporting Initiative (GRI). 2005. Global Reporting Initiative homepage. http://www.globalreporting.org.

Hart, S. 1997. Beyond greening: Strategies for a sustainable world. *Harvard Business Review* (January-February): 66-77.

Hawken, P. 1993. *The ecology of commerce: A declaration of sustainability.* New York: Harper.

Hawken, P., A. Lovins, and H. Lovins. 1999. *Natural capitalism: Creating the next industrial revolution.* Boston: Little, Brown.

Holliday, C., S. Schmidheiny, and P. Watts. 2002. *Walking the talk: The business case for sustainable development.* Sheffield, UK: Greenleaf.

KPMG. 2005. *International survey of corporate sustainability reporting.* http://www.kpmg.ca/en/industries/enr/energy/documents/KPMGCRSurvey.pdf.

Lafferty, W., and O. Langhelle. 1999. Sustainable development as concept and norm. In *Towards sustainable development: On the goals of development and the conditions of sustainability,* ed. W. Lafferty and O. Langhelle, ch. 1. London: Macmillan.

Lafferty, W., and J. Meadowcroft, eds. 2000. *Implementing sustainable development: Strategies and initiatives in high consumption societies.* Oxford: Oxford University Press.

Lowe, E. 1996. Industrial ecology: A context for design and decision. In *Design for Environment: Creating eco-efficient products and processes,* ed. J. Fiskel, ch. 26. New York: McGraw-Hill.

McDonough, W., and M. Braungart. 1998. The next industrial revolution. *Atlantic Monthly* (October): 82-92.

–. 2002. *Cradle to cradle: Remaking the way we make things.* New York: North Point Press.

Mining Association of Canada. 2005. *Towards sustainable mining.* http://www.mining.ca/english/tsm/tsm-eng.html.

National Academy of Engineering. 1999. *Industrial environmental metrics: Challenges and opportunities.* Washington, DC: National Academy Press.

National Roundtable on the Environment and the Economy (NRTEE). 2003. *Environment and sustainable development indicators for Canada.* Ottawa: NRTEE.

PricewaterhouseCoopers. 2002. 2002 sustainability survey report. http://www.pwc.com/fr/pwc_pdf/pwc_sustainability.pdf.

Rees, W. 1995. Reducing the ecological footprint of consumption. Mimeo.

Robèrt, K-H. 2002. *The Natural Step story: Seeding a quiet revolution.* Gabriola Island, BC: New Society Publishers.

Roodman, D.M. 1999. Building a sustainable society. In *State of the world,* ed. L. Brown, C. Flavin, and H. French, 169-95. New York: Norton.

Roome, N., ed. 1998. *Sustainability strategies for industry: The future of corporate practice.* Washington, DC: Island Press.

Schmidheiny, S., with the World Business Council for Sustainable Development. 1992. *Changing course: A global business perspective on development and the environment.* Cambridge, MA: MIT Press.

Stora Enso. 2003. *Environmental stewardship report 2002.* Duluth, MN: Stora Enso North America.

Stratos Inc. 2005. *Gaining momentum: Corporate sustainability reporting in Canada.* http://www.stratos-sts.com.

United Nations Conference on Environment and Development (UNCED). 1992. *Agenda 21.* Rio de Janeiro: United Nations.

United Nations Environment Programme (UNEP). 2005. *Living beyond our means: Natural assets and human well-being.* New York: UNEP.

United Nations Environment Programme (UNEP) and World Business Council for Sustainable Development (WBCSD). 1997. *Eco-efficiency and cleaner production: Charting the course to sustainability.* Paris: UNEP.

Vector Research. 2001. Opinion poll. In *The new balance sheet: Corporate profits and responsibility in the 21st century,* Canadian Democracy and Corporate Accountability Commission. Toronto: CDCAC.

von Weizsäcker, E., A. Lovins, and H. Lovins. 1998. *Factor four: Doubling wealth, halving resource use.* London: Earthscan.

Wackernagel, M., and W. Rees. 1996. *Our ecological footprint: Reducing human impact on Earth.* Gabriola Island, BC: New Society Publishers.

White, R. 1994. *Greening of industrial ecosystems.* Washington, DC: National Academy of Engineering.

World Business Council for Sustainable Development (WBCSD). 2000. *Eco-efficiency: Creating more value with less impact.* Geneva: WBCSD.

World Commission on Environment and Development (WCED). 1987. *Our common future.* New York: Oxford University Press.

Xerox. 2004. *Environment, health and safety progress report.* http://www.xerox.com/environment.

3
Policy Instruments and Sustainable Production: Toward Foresight without Foreclosure

Robert Paehlke

Sustainable production is "dematerialized" production. That is, it is production that creates economic outputs efficiently in terms of energy and material extractions and throughputs and sees societal well-being as an end goal with GDP as an intermediate goal (for full development of this perspective, see Paehlke 2003, especially Chapters 4 and 5). Reduced extractions and throughputs – a narrower meaning of sustainability – encourage, but may not in themselves guarantee, clean production and production that is respectful of the space of non-human species. Central to "narrow" sustainable production are product durability, the use of waste streams as raw materials, energy efficiency, product and packaging redesign, the substitution of communication for transportation (as with Internet-based financial services or telecommuting), and the supplanting of physical goods by knowledge goods. "Broad" sustainable production continuously and simultaneously shows measurable improvements in environmental quality, reduced impacts upon non-human habitat, increased economic output per unit of energy and material input, and improving societal well-being.

Resource sustainability could also be achieved through reductions in economic output and/or well-being, a less desirable option for obvious reasons. Getting ahead of the curve on sustainable production enhances competitiveness in almost every plausible economic scenario. David McGuinty argued that as global carrying capacity for all species, including humans, nears its limit, investments in the environment are tantamount to long-term investments in our economy (McGuinty 2000, 3). Another way to put this is to say that investments in sustainable production extend the range of our economic possibilities into the future. Moreover, economies that are especially innovative in this regard keep open options that will be foreclosed elsewhere. Keeping those options open is doubly economically beneficial because today's saved resources will be more valuable in the future and because the efficient technologies are themselves exportable.

At first blush, Canada would not seem a likely location for radical innovations in sustainable production. Historically, and even today, we are a nation particularly blessed with resources and singularly adept at resource extraction. Few Canadians think about resource shortfalls other than perhaps in ocean fisheries; we more often think about how to promote increased global resource use. We need to better understand, however, that to the extent that Canada can innovate in sustainable production, we can gain twice: we can lower the cost of our manufactured goods and either free resources for export or enable domestic industrial use of those resources to continue for longer (so long as they are not replaced by new technologies, an unlikely scenario in the case of fossil fuels, for example). Indeed, we might gain a third time through the opportunity to export sustainable production processes, technologies, and designs. Moreover, it is clear that in some cases resources have been (and are being) overused in Canada; others are in increasingly short supply in North America as a whole (fresh water, electricity, and natural gas, for example); and, most crucially, conventional fossil energy production will soon peak (and given climate change considerations and the range of environmental concerns associated with non-conventional energy sources, this type of energy should be used more sparingly in any case) (*Scientific American* 1998).

Rising energy prices make sustainable production all the more essential. Virtually every resource input is energy-intensive in its extraction and processing. If energy prices are high, fibre, metals, chemicals, and the transportation of both raw materials and finished goods become more expensive. In this context, increased output, and indeed improved societal well-being (a term that will be elaborated more clearly below), become ever more dependent on the development of sustainable production, requiring ever more economic output per unit of input of energy and materials – in a word, dematerialization. Gradual dematerialization is, of course, a long-standing trend in industrial and post-industrial societies. For more than a century, GDP has grown on average more rapidly than have energy and material inputs, but extraction and energy use have nonetheless risen in synch with, though more slowly than, economic growth. Ultimately in many cases, and in some cases soon, inputs will be capped (plausibly fibre and, in many locations, fresh water), and in some cases (fossil fuels and plausibly energy use itself) total global consumption will contract. The challenge, especially for advanced economies, is to increase the rate of dematerialization now. It is not, in my view, too ambitious a target to seek in the short term a rate of dematerialization, especially regarding energy use, equal to the average rate of economic growth. All other environmental problems aside, the peaking of fossil fuel supplies and the extreme poverty of half of the world's population would seem to make this a minimally prudent goal (Wackernagel and Rees 1996; Carley and Spapens 1998).

Defining Sustainability and Sustainable Production

As noted, sustainable production can be defined either broadly or narrowly. Broadly, one might identify environmental sustainability as producing the necessities of a quality human existence on a continuous basis within the bounds of a natural world of undiminished quality. Environmental sustainability encompasses high performance levels on each of the three value realms of environmentalism: (1) environmental health (air and water quality, and clean production); (2) the protection of habitat, wilderness, ecology, and biodiversity; and (3) resource sustainability. Resource sustainability, the narrower goal, is essentially the effective management of society's total material (and energy) requirements (TMR). Sustainable TMR involves continuous improvement in the material and energy extracted from nature per unit of economic output. Ideally we would find ways to reduce the total quantity of material and energy consumed by an economy. In any case, sustainable TMR involves minimizing extraction and maximizing materials reuse and recycling, as well as limiting the use of renewable resources to amounts well within nature's capacity to supply them in perpetuity.

Robinson and Tinker (1997) identify two forms of "decoupling" as essential to moving toward sustainability. One is the decoupling of economic output (GDP) from energy and material throughputs, especially the extraction of raw resources from nature. The other is the partial decoupling of social well-being from GDP per capita (that is, improving quality of life faster than increases in economic output, or getting more for our money). While the former might be achieved through, for example, green taxation or subsidies to sustainable design and process innovation, the policy basis for achieving the latter is less obvious and likely more politically controversial. The decoupling of prosperity (taken as total consumption) from well-being might be achieved, for example, through reduced crime rates, improved health outcomes, or educational quality – or even through increased leisure. Many such shifts are highly value laden (and may connect closely to changes in economic distribution), but some are related to sustainable production on a societal scale. Some examples of this (such as changes in urban form and improved transportation systems) will be considered below. Generally, however, this chapter will emphasize the first form of decoupling, since arguably the second is primarily achieved outside of the production process.

The overall logic of all of the means for advancing sustainability becomes clearer within the three-dimensional model for evaluating societal performance in terms of economic prosperity, social well-being, and societal "metabolism" (Fischer-Kowalski and Haberl 1998). Societal metabolism is the rate at which materials and energy are extracted from and returned to nature. Within this model, each of these dimensions is understood and measured separately and independently. It might be argued that societal well-being (as measured by health, education, comfort, happiness, social cohesion, or any number of indicators other than wealth and income) is the primary objective and that it is achieved through prosperity, which in turn depends on the amount of materials and energy that are extracted from nature. These causal links can also be seen as the points at which partial decoupling can be achieved, essentially through increases in productive efficiency (in aspects of efficiency in terms other than labour per unit of output). It is a mistake to simply assume that the only way to improve well-being is to improve prosperity, just as it is an error to assume that the only way to enhance prosperity is to increase extraction and throughputs. Both assumptions simply miss the point regarding sustainability.

This perspective distinguishes our model from a "common sense" model that takes prosperity (as measured by GDP) to be the essential goal of society and the overwhelmingly dominant objective of public policy. This latter model might be called "economism," and from the perspective of sustainability analysis might be said to consistently misdirect public policy by misconceiving fundamental objectives. There are three bottom lines, not one; economism is unduly one-dimensional (Paehlke 2003). The mere fact that greater prosperity *could be* channelled to environmental and social improvements is irrelevant. Again, each of the three objectives (well-being, prosperity, and resource or environmental sustainability) must be measured independently. Rising wealth is better than declining wealth, but at least as important are two other questions: is social well-being rising both absolutely and in terms of efficiency of production per unit of prosperity, and are we increasing materials and energy efficiency (GDP/TMR) at least as fast as we are expanding GDP? Many would argue that such a rate of improvement in societal metabolism is insufficient to the achievement of sustainability, but all would agree that it is a necessary interim goal (Carley and Spapens 1998; von Weizsäcker et al. 1998).

Definitions are crucial to the achievement of sustainable production. We must be clear about what it is we are trying to sustain and how it is that we will know when and if we are making progress. The achievement of broad sustainability involves sustaining our economy and the resources on which it depends, sustaining the quality of human life and therefore non-toxic air, water, and land, as well as a quality climate, and sustaining biodiverse wild nature (both for its own sake and to sustain both resources and the quality of human life). Every material input extracted from nature impacts one or more of these objectives. The reduction of, or slower growth in, TMR is thereby a rough measure of the extent to which sustainability is being achieved. The great challenge is to limit TMR while avoiding recessions or economic decline. This form of material flow analysis, as its practitioners acknowledge, also needs to be adjusted for the great variability in associated environmental impacts (Matthews et al. 2000). Once such calculations are standardized, this measure (TMR) can provide a measure of sustainable production within national economies, industrial sectors, or firms.

It is also important to remember that sustainability encompasses all aspects of the economy, both production and consumption. This is different from a concern with pollution, which often results from a limited number of processes or substances within particular industries, and which can be ameliorated with a limited number of innovations or interventions. Government nonetheless has a role to play in sustainable production. The market alone will not necessarily establish economic parameters that motivate and foster a sufficient level of sustainability innovations. It is essentially a matter of getting resource prices right – less a matter of competition among economically desperate nations than of all nations acting cooperatively to maximize the economic, social, and environmental dimensions of our collective future.

The above is a succinct statement of the theoretical perspective of this chapter. The remainder of the chapter will consider examples of notable sustainable production opportunities. Detailed quantitative analyses will not be undertaken, but this brief look at some technological and socially innovative possibilities, including discussion of some sustainability risks and failures, could be used to select policy instruments that promote opportunities without foreclosing in advance innovations that are difficult to assess. Technological innovations noted will include the potential of the telecommunications sector and the possibly more controversial potential interconnections among agro-forestry, biotechnology, and climate change. New telecommunications technologies, though they themselves consume considerable energy, could lead to greater overall energy and materials efficiency. Likewise, while the genetic modification of plants may pose new ecological risks, they might also permit enhanced carbon sequestration. We will consider how, in these complex cases, we might make balanced and appropriate choices. The risks and failures discussed will include energy consumption per unit of GDP, urban sprawl, forest extraction, and fisheries and agriculture.

This chapter then considers the central challenge in risk decisions and policy instrument choice – how do we gain sufficient foresight to intervene in terms of risk management without foreclosing economic and technological innovation? Improved risk decisions involve many things, including the need for multi-dimensional and comparative analyses, more effective and visible social and sustainability indicators, increased public involvement in risk decisions, and a wider recognition that products evaluations cannot be isolated from the processes within which they are produced and requires a caution regarding self-interested science. Instrument choice may well centre on achieving global prices for energy and raw materials that take into account the long-term ecological costs and sustainability implications of the present trajectory of the global economy. Innovation in sustainable production results from the intelligent anticipation of future trends and requires a climate that is likely to reward innovation. There are policy instruments that would achieve these ends, at least in part. We will consider some of these possibilities following a brief review of the kinds of economic shifts and sustainable production opportunities that might exist within and between several key economic sectors.

Telecommunications and Sustainable Production

One of the sectors within and through which sustainable production can be enhanced is telecommunications, broadly defined. While telecommunications requires energy and electricity consumption, many of the products of this sector are highly dematerialized relative to other economic sectors, and could be more so. Computer software, for example, involves very high economic value and many high-paying jobs per unit of energy and material input. This ratio is improved by the elimination of packaging and shipping when software products are preinstalled or downloaded from the Internet. These changes only begin to scratch the surface of possibility within this sector, however. The greatest potential lies in the use of computers and telecommunications to substitute knowledge products for material products, to replace transportation with communications, to alter the relationship between home and work locations, and to automate industrial and commercial activities. The first two of these would seem to contribute in a straightforward way to a sustainable society; the latter two carry risks as well as positive potential.

Money spent on downloaded video games cannot be spent on new snowmobiles, making consumption sharply less energy- and materials-intensive (even if one assumed that the computer had no other function). Even something as simple as online access to news stories involves the use of less energy and materials than the purchase of a daily newspaper, especially if one drove to a store especially to purchase that paper. While obviously the paperless office is not imminent, or even probable, it is not impossible that some uses of paper could be curtailed without significant loss of amenity if paper prices were to rise owing to alternative uses for paper's material inputs, such as for the development of alcohol from biomass as a future energy source. A third possibility for knowledge/materials substitution lies within the world of telecommunications, in the evolution of wireless technologies; cell phone systems are far less materials-intensive than wired phone systems.

The potential for more sustainable production within every industry and commercial realm can be seen in the increasing possibilities for substituting communications for transportation. In a realm familiar to many of us, webcast university courses are now a reality and certainly have the potential to greatly reduce the transportation needs of participants. Teleconferences have comparable, if not greater, potential in this regard. In both cases, it is of course arguable that something is lost in terms of the quality and effectiveness of the interaction. Such substitutions, however, could be partial rather than complete. Alternate annual conferences could be virtual, or national or global conferences could be virtual and regional conferences actual. Lectures could be virtual and some seminars actual. The realm of communications-for-transportation substitutions is enormously larger than this, however, and includes: (1) e-mail attachments in lieu of courier services; (2) worldwide delivery of films to theatres via satellite (rather than physical shipping); (3) interactive television in lieu of trips to and from the video store; and (4) electronic financial management (as well as automated teller machines in locations that one would have gone to in any case). Each of these changes is significantly less energy- and materials-intensive.

Telecommuting is another widely touted possibility with very significant potential sustainability gains. Those gains derive from reduced needs for transportation and for office space (spaces could be shared by employees who come to the workspace on different days). One sustainability risk in this is that an irregular need for commuting might lead to an increase in the average distance travelled. The sustainability effects of telecommuting depend very much on the manner in which, and the extent to which, it develops, but even an arrangement where a proportion of employees work from home for one or two days a week should reduce the amount of travel. In general, the amount of telecommuting done at present is underestimated because statistics may exclude those who work as consultants and contract workers, a group that has grown in recent years.

Automation could have more significant sustainability implications than telecommuting. Automated production or distribution facilities can operate 24 hours a day and 365 days a year. Far less space is thus needed for production, and fewer employees will travel back and forth to work. Moreover, if a facility is fully automated, it need not be heated or lit to the same standard as a human-centred workspace. All of these changes have potentially positive sustainability implications. The risk is that more goods will be produced at lower cost, resulting in increased production and an increased use of raw materials and energy. This risk would be reduced if some of the productivity gains resulted in additional leisure rather than additional output (Paehlke 2003, Chapter 7). The sustainability implications also depend on the kind of industry that is automated. Nonetheless, it is possible that automation could eventually result in a transformation of lifelong patterns of work and consumption that would be more sustainable.

The most significant sustainability cost within the electronics sector as a whole is the rapidity of technological obsolescence. There is only a limited second-hand market for older computers, but this could be enhanced through domestic or international charitable contributions, thereby reducing (or at least delaying) wasted material. Electronic wastes frequently contain toxic components and are becoming an increasingly serious environmental problem (Ross 2005; *New York Times* 2005). Thus the better solution lies in design improvements and the reduction or elimination of electronic wastes through design innovations and recycling. One small change has already been made – monitors and keyboards are now sold separately and can be kept as computers are replaced. A great deal more could be done, however. For example, it could become normal practice to replace a chip or a drive while keeping everything else intact, thereby reducing wasted material by several orders of magnitude. Combined with improved energy efficiency, the industry could continue to show strong improvements in dollar value per unit of material throughput. Nonetheless, the computer and telecommunications sector as a whole adds significantly to the size of the economy without adding commensurately and proportionately to overall energy and material throughputs. This is also true for the health care sector, although the pharmaceutical industry has a number of associated pollution problems, including endocrine disruptors (Colburn et al. 1996). Low energy and material throughputs are also the norm in education, financial services, and the public sector. These economic sectors account for a good share of the growth potential of advanced economies, and to the extent that they account for an increasing proportion of the economy by supplanting other activities, they advance the sustainability of the whole society.

Canada, however, still lags somewhat in terms of the development of these sectors, and this lag is not unrelated to government policies. The Organisation for Economic Co-operation and Development (OECD) country report for 2000 makes two relevant points here. First, "in Canada, the two industries which are leaders in the 'new economy' – electronic equipment and industrial machinery – make up a smaller share of output and have made slower productivity advances than in the United States" (OECD 2000, 12). More important, however, were the OECD's extensive comments on Canada's reluctance to limit disproportionate subsidies to economic sectors at the opposite end of sustainability (OECD 2000, 17):

> The sense of "immensity" and Canada's rich endowment of natural resources have led to policies favouring their development and use ... Resource-based production is energy-intensive and may in the end cause serious pollution problems, notwithstanding the Canadian environment's substantial assimilative capacity compared with other OECD countries. Support has been

> especially important for activities based on non-renewable resources (such as oil, gas, metals and minerals), coming in the form of preferential tax treatment. Although the recently announced tax measures will, over time, contribute to leveling the playing field, this has put other sectors of the economy, such as knowledge-based industries, at a disadvantage.

These comments are altogether in keeping with the sustainability-oriented analysis of environmentally oriented research organizations (Roodman 1996; Durning and Bauman 1998). It comes, however, from a leading bastion of economic orthodoxy. In green jargon, Canada would be said to be continuing to promote a cowboy economy, an approach to economic development that presumes an endless frontier. Even if resource and assimilative capacity remains in Canada, it is time to recognize that in a globalized economy one gains little by behaving in a manner out of line with the resource and environmental realities of the world as a whole. Resources such as energy, water, and high-grade wood will not lose value over time. Subsidizing their extraction makes no sense either in terms of sustainability or in terms of the long-term health of the Canadian economy.

Sustainable Production, Transportation, and Urban Form

In its volume on sustainability, the World Resources Institute selected transportation as one of four US economic sectors for study (the other three were electricity, agriculture, and forestry) (Dower et al. 1997). This seminal work speaks of three phases of environmental history in the United States: first, resource exploitation and unchecked pollution, followed by a phase now ending that centred on pollution abatement, and now a third phase in which the transportation sector features prominently. The authors describe the third phase in this way: "In the third more systematic changes will be needed. It's no longer enough to fix economic development's environmental side effects. US policies must tackle underlying causes, not just the symptoms. This means preventing, rather than treating, industrial pollution. It means designing whole new systems for moving people, goods and ideas" (Dower et al. 1997, 1-2). Other possible initiatives are then listed, but transportation policy is emphasized second only to pollution prevention. The authors go on to state that "our petroleum-based ground transportation system ... is threatened by the long-term depletion of global oil reserves" (3). The North American transportation system is the least sustainable in the world.

In terms of energy use, buses and trains are more efficient (in energy use per passenger mile) than either automobiles or airplanes. Airplanes are especially fuel-inefficient on short-haul flights under 800 kilometres (since most fuel is used in taking off and landing). Yet in the United States some 95 percent of passenger miles travelled are by car, truck, or plane. There are three motor vehicles licensed for every four Americans, excluding virtually only the young, the very old, and some of the urban poor. Canada has a marginally more sustainable transportation system, especially within the urban core of major cities, although many US cities are now taking significant initiatives (Portnoy 2003). In Toronto (as distinct from the Greater Toronto Area, or GTA), for example, about 30 percent of the trips to work are not by automobile. Even in Greater Toronto as a whole, however, the percentage of trips on public transportation is declining and urban sprawl is accelerating. In North America, our trillion-dollar transportation system (20 percent of GDP) is less sustainable now than it was in 1990.

Transportation cannot be considered in isolation from other sectors of the economy and governance – especially construction and urban planning. The choice of transportation mode within any given urban area is a function of urban form and design. Using data from a study of thirty world cities, Newman and Kenworthy (1999) found that residential density, combined with mixed-use planning, was a clear determinant of choice of transportation mode. That is, when cities are relatively compact (a mix of multiple-family dwellings and single-family houses with relatively narrow lots arrayed on square street grids) and commercial and other non-residential functions are nearby, people will opt for public transportation in fairly significant numbers. This generalization seems to hold regardless of other cultural and economic factors (the cities in the study were in Australia, Asia, Europe, Canada, and the United States). This is one reason why environmentalists are so tenaciously opposed to urban sprawl. Another is that sprawl has massive effects on habitat and biodiversity, as does the automobile (which requires a large multiple of the land used by public transit to transport an equal number of people).

The keys to sustainable transportation, however, are the modal shifts and reduced daily mileage of automobile use that are associated with more compact and mixed-use urban forms. In US cities, gasoline consumption averaged 58,541 megajoules per person (Paehlke 1991, I-1). Toronto (Canada's best performer) was at a level of 34,814, but European cities (including Stockholm, London, Paris, and Frankfurt) averaged 13,280. City by city, the proportion of trips to work by public transit, walking, or cycling, as well as gasoline consumption per capita, was a linear function of overall urban density. The relative price of driving and public transit can alter this equation somewhat, as in the case of switches from employer-provided parking to pay-for-parking arrangements (or changes in the level of subsidies to urban transportation). To the credit of the current Ontario government, the recent initiative to offer public support for housing construction within the urban core of Toronto has considerable potential. Better would be the establishment of minimum overall residential density requirements within the 905 area code communities and their equivalent in other Canadian urban areas.

Sustainable production, it should be clear by now, is a comprehensive and complex objective. It is not simply a matter of industrial product and process design. One must include the energy and materials involved in transporting employees, raw materials, manufactured components, and finished products. Nonetheless, there is also enormous scope for product design where automobiles, public transportation systems, transportation (and other) infrastructure components, building materials and contents, and buildings themselves are concerned. Indeed, taken as a group, these products comprise a significant proportion of all of the extractions and throughputs in a developed economy. Building materials, an economic sector in which Canada is a major player, can involve up to 10 to 12 tons of material per person per year (Matthews et al. 2000, 71). In the United States, for example, construction and transportation accounted for over 20 percent of all extractions and throughputs (Matthews et al. 2000, 25).

A decision to infill housing and commercial development within an urban core where existing infrastructure can be utilized greatly reduces throughputs and thereby enhances sustainability. Moreover, the number of people or businesses served by a given amount of sidewalk or storm sewer or telephone wire is increased because distances are decreased compared with suburban building arrays. Restoration and conversion from one use to another (rather than new construction) results in additional savings. In addition, building materials can be lighter and/or produced from recycled feedstock. For example: (1) more durable asphalt roads have been produced using material recovered from old tires; (2) gravel (a large component of concrete production) can be replaced by construction waste; (3) countertops and other products can be manufactured from waste paper; and (4) plastic substitutes for pressure-treated wood products can be produced from recycled materials. In every case here, the result is enhanced sustainability and reduction in environmental impacts. A comparative study of the relative environmental impacts of basic building materials (steel, concrete, and wood) reveals that all have significant impacts, suggesting that building adaptation and reuse is almost always the best environmental option (Paehlke 1994; von Weizsäcker et al. 1998, 78-80).

We have known since the energy crisis of the 1970s that energy use in buildings can be radically reduced with new designs. Model houses that achieved radical reductions were built in Saskatchewan in the mid-1970s and I have been in several in and around Peterborough, Ontario, that date back that far, including one south-facing, partially buried house that is heated with almost no energy. With the decline of energy prices around 1985, virtually all of these efforts faded. Nonetheless, some highly efficient buildings have been constructed, including, for example, the headquarters of the Rocky Mountain Institute near Aspen, Colorado, and office buildings in Sweden (cold climate), Britain, and inland California (warm climate). All achieve energy reductions of at least 75 percent using insulation, passive solar, thermal windows, and various economically sound designs (von Weizsäcker et al. 1998, 10-29). One of the best is the large (55,000-square-metre) ING bank building in Amsterdam, which uses 20 percent of the energy of comparable new office buildings at an additional cost for energy efficiency repaid within three months in energy savings. We know, technically, what to do, but typically do not do it primarily because long-term operating costs (and "external" sustainability and environmental issues) are not yet central to all design, corporate, or governmental decision making.

Thus, construction is important to sustainable production in three ways: (1) energy efficiency in building operation, (2) construction materials sustainability, and (3) the fact that urban form largely determines transportation sustainability. It is especially important to recognize that there is more to sustainable transportation than enhanced automobile fuel efficiency, as important as that is. First and foremost, governments must create the conditions to ensure that both manufacturers and consumers will move rapidly to more sustainable transportation options. This would include both maximizing automobile fuel efficiency and developing and aggressively promoting effective public transportation systems.

Canada generally, and Ontario and Toronto in particular, have done the opposite to a point now where the proportion of Toronto Transportation Commission (TTC) income that comes from the fare box is the highest share of any large city in the developed world. Virtually all highways are built and maintained with funds from the public treasury. Given the environmental advantages of public transit, either many more roads should be toll roads or the share of public revenues devoted to transit should be increased significantly. Such initiatives could be enacted with imagination as in Portland, Oregon, where within the downtown core streetcars are free, thus relieving busy streets of excessive traffic and encouraging the use of park and ride options.

As the OECD (2000, 19-20) put it regarding Canada's Kyoto obligations (and therefore also, in the broad view, sustainable production):

> Even if Canada is able to buy GHG emission quotas on an international market, it will probably have to take steps to accelerate the reduction in domestic fossil-fuel consumption per unit of GDP. In this case, rather than resort to command-and-control-type regulations, it would be advisable to rely primarily on a cost-effective instrument, such as a tradable permit scheme, and not to exclude specific sectors (such as transport and energy) from its application. Increased taxes on fuel might be helpful to reduce emissions related to transport. In any case, measures will need to be implemented well before the target period in order to allow for a gradual adjustment in the energy-using capital stock.

It is crucial to select policy instruments that advance all aspects of sustainable production, from sectoral shifts within the economy (as from resource extraction to knowledge-based sectors), to modal shifts in transportation, to improved efficiency of energy use in vehicles and appliances. The efficient use of energy within the transportation and energy sectors are an important part of the sustainability of all production, since these, like food and housing, are deeply embedded within the whole of the economy.

Also critical is an appreciation of the importance of lag time in a gradual adjustment in the energy-using capital stock. Within the consumer realm, the stock of automobiles turns over in a period of about ten years, but bringing a qualitatively different vehicle on-stream takes an additional five years or more. Large appliances may only turn over in a period of twenty or thirty years. Interesting, however, within this latter example is the case of refrigerators. As Boyd (2001) reports: "As recently as 1975, the average North American refrigerator used 1,800 kilowatt hours (kWh) per year. As of 2001, the latest refrigerators use 75 percent less energy, without a decrease in performance or significant price increase." This sustainable production initiative, which is still a long way from having fully worked its way into the North American stock and is still short of what is technically possible, was spurred by a combination of regulations and design incentives to manufacturers (from US utilities). An even longer time lag is associated with municipal infrastructure, buildings, and the overall shape of cities; here, the stock in some cases will last for centuries. Sustainable production is often not a quick fix, and production incentives and disincentives should thus anticipate future sustainability realities – no easy matter.

Arguably, the key to anticipation is energy efficiency. This is not, however, just a matter of fuel efficiency, or in the realm of transportation, modal choice. Recalling the example of asphalt produced from scrap tires, the result is not "just" an environmental saving within landfill sites. The tire material replaces material from a fossil fuel source. In the case of BMW automobiles designed to be almost wholly recyclable, the result is not just fewer rusting autos but less energy involved in the "re-production" of the next car. There is also an enormous reduction in air pollution and energy use when the design change eliminates the need to burn off some plastic parts in the process of recycling the metal within the automobile body. In some cases, switching to metal parts from plastic parts could enhance recyclability, but would add to the weight of the vehicle and thereby reduce fuel efficiency. Understanding how best to proceed requires doing careful life cycle assessments.

Finally, under the heading of transportation and urban form is the possibility of auto-sharing cooperatives or firms. This European initiative came to North America only recently and is now underway in some Canadian cities,[1] as are some excellent carpooling initiatives. Auto-sharing cooperatives are an option that is midway between ownership and car rental. Participants make an equity investment in a neighbourhood vehicle fleet and then pay a relatively small amount on a per-use and per-kilometre basis. This particular type of initiative suggests the breadth of meaning that may be attached to "production" within sustainable production. Auto-sharing coops and firms "produce" assured access and shared ownership of automobiles rather than a physical product. Sustainability is enhanced in that fewer vehicles are produced as each vehicle serves on average five to eight owners. Yet individuals still have a car when they need it (for the daycare carpool, heavy shopping, or weekends out of town). Users otherwise depend on bicycles or public transit. Overall, in one closely studied European example, fuel use was reduced by a factor of two and vehicle production by a factor of four (von Weizsäcker et al. 1998, 129-30).

All of the above examples apply throughout the wealthy nations, but may be especially useful within poorer nations, where automobile-dependent infrastructure patterns are not so fully fixed. The next section considers an economic sector of particular significance to Canada, both in terms of the present structure of our economy and in terms of potentials for the future.

Sustainable Production: Forests, Farms, and Fibre

Agriculture and forestry have always been a large part of the Canadian economic experience. And, given our very large land base and an almost inevitable growth in demand for food and fibre, these sectors will almost certainly be a large part of Canada's economic future. There are very real threats to the sustainability of these sectors, but there are also many very significant sustainable production opportunities. The threats include overfishing, overcutting of forests, soil erosion associated with agriculture and forestry, climate change, farmland losses (primarily to urbanization), surface and groundwater pollution, and losses to the genetic diversity of crop plants and domesticated animals. The opportunities include the development, within these sectors themselves, of means to ameliorate or reverse environmental threats. Specific opportunities include the use of our land base to reduce the risk of climate warming through the incremental sequestration of carbon in soil, crops, or forests, even perhaps using genetically modified organisms to this end.

Faeth (1997, 54-55) ranks genetic loss and climate change as the greatest risks to agricultural sustainability, followed by nutrient runoff and pesticide pollution. Regarding genetic losses, he argues that "the extinction of genetic stocks irretrievably closes off future opportunities for crop and livestock improvement and the development of genetic responses to emerging pest and climate threats. Genetic extinction is forever and it represents an unknowable but undeniably large opportunity cost" (56). This (even though written in 1997) appears to exclude the possibility of molecular genetic manipulation, but remains a largely valid assertion since many other adaptation possibilities are excluded nonetheless. Regarding climate change, Faeth contends that "no other issue" could "affect agriculture at so large a scale," and he therefore ranks climate change as a large threat to sustainability. If climate change is more rapid it will create a greater challenge to agriculture and lessen the prospects for effective adaptation. We simply do not know, however, how rapidly change in either average temperature or precipitation will occur or what the particular effects in any given location will be.

Faeth (1997, 62) also reports that while other economic sectors have made gains in emissions reductions, agriculture "has become the main source of water quality impairment." In Canada, recent concerns about large-scale animal operations have raised alarm bells throughout the country, as have concerns regarding the risks posed to groundwater by large numbers of abandoned and uncapped agricultural wells. A highly publicized and fatal water contamination event in Walkerton, Ontario, have accelerated such concerns. The massive increase in the scale of feedlot operations has, in effect, moved industrial operations into rural areas. This problem is even more pronounced in the United States, where one poultry operation in Arkansas produces millions of chickens per day and hog farms in the Carolinas resulted in widespread water contamination during floods in the late 1990s.

Forestry practices, however, have probably been the most controversial environmental sustainability issue in Canada. In British Columbia, there have been challenges to further reductions in old-growth habitat conditions, to the loss of spawning grounds for salmon (lost as forest cutting creates siltation and/or temperature increases), to erosion of hillsides, and to non-sustainable cutting rates generally (Wilson 1998). In particular, there have been suggestions that the current rate of cutting will remove all non-protected forests in BC before sufficient second-growth forests are available. Ninety percent of forest harvests in Canada are sold on the export market, mostly to the US, where the area of legally protected forests has been expanded from 20 million acres in 1960 to 47 million acres in 1992. Former President Clinton further expanded protections as he left office, but most of his initiatives and some previous protections have been systematically eroded by the administration of George W. Bush. (For details regarding the Bush rollbacks, see Cousins et al. 2005.) The irony in this is that Bush is an unintended friend of sustainable forest practices in Canada.

On the other hand, during roughly the same period, recycled fibre went from 8 percent of the feedstock of the US pulp and paper industry to 33 percent. The bottom line in all this, however, is this: "If current trends in wood-products consumption and net annual timber growth continue over the next decade, the United States will consume more wood fiber than it grows in its timber base" (Johnson and Ditz 1997, 193). Given that forest growth is, on average, slower in Canada than in the US (owing to differences in climate), and the rapid development of extraction operations across the northern tier of Alberta, Saskatchewan, Manitoba, and Ontario, it is likely that this will soon also be true of Canada. The sustainability of the forest industry in North America, at present rates of cutting and growth, is far from assured. Moreover, present levels of wild forest habitat will likely decline since second-growth forests are typically far less biologically diverse. The greater danger, however, is that additional pressures will be placed on our forest resources in the future.

One reason for the latter possibility is that as fossil energy supplies decline and energy prices rise, wood may look better as a building material than either steel or concrete, both of which are significantly more energy-intensive products. The larger reason, however, is that the output of our forests and farms is a potential source of alcohol-based fuels for motor vehicles. Initial indications of this possibility were put forward in an article by US Senator Richard Lugar and former Central Intelligence Agency director James Woolsey in the influential journal *Foreign Affairs:* "Recent and prospective breakthroughs in genetic engineering and processing ... are radically changing the viability of ethanol as a transportation fuel. New biocatalysts – genetically engineered enzymes, yeasts and bacteria – are making it possible to use virtually any plant or plant product (known as cellulosic biomass) to produce ethanol" (Lugar and Woolsey 1999, 89). The authors go on to make the economic and geopolitical case for the radical expansion of ethanol production from plant biomass.

The good news in terms of sustainability is that this production could come in part from agricultural wastes such as grain stalks, from scrap paper, and from corncobs. The perhaps not-so-good news is that additional demand for plant matter might put additional pressures on already burdened forests. Lugar and Woolsey mention agricultural field-based "wastes" but do not explore either forest or other reuse possibilities. It is arguable, as well, that agricultural by-products should be returned to the soil. Interestingly, in the mid-1970s environmentalists in Canada made a case for the extensive use of forest-based alcohol fuels as a renewable (we would now say sustainable) alternative to fossil fuels (Lovins et al. 1979). To the extent that such an initiative involves afforestation, putting land back into forest production that is now low-grade pasture or agricultural land, this should result in increased carbon sequestration. To the extent that it involves additional forest removals, it would likely be negative in this regard and could render our existing forests less sustainable. Anything that extends the life of fossil energy supplies should be seen as a plus, however, and the environmental effects of substitutes should be seen in relation to fossil fuel impacts, not against an idealistic standard of no effect.

There is also the possibility of developing genetically modified trees that would either grow faster (thereby sequestering more carbon) or be altered so that tree cellulose might be more readily converted to ethanol, or both.

In nature, of course, as the cliché would have it, there is no free lunch. Faster-growing trees would take up additional soil nutrients and more water, and thus could not be grown as successive crops indefinitely without adequate rainfall and some addition of nutrients. Moreover, fast-growing poplar monocultures would not necessarily provide as rich a habitat as the forest that they would replace. These and other effects will need to be explored, as will the highly complex ethical and sustainability questions associated with the new science of genetic modification. We do not know all of the possible ecological risks of such an undertaking. This is not, however, an irrefutable argument against proceeding so much as a caution to proceed with great care.

One risk is that modified plant (and animal) species could come to replace "natural" plant (and animal) species. This may seem implausible, but the concern of Faeth regarding genetic losses in crop species mentioned above would also have seemed far-fetched but a century ago. That is, we humans are now so dominant a species that little grows in agricultural (and increasingly forest) settings that we have not planted or bred deliberately. "Wild" nature has selected the array of species that exist over countless eons of highly variable conditions. It would just seem right in principle that such species should not be comprehensively supplanted by "untested" human-created species. We humans do not, and never will, understand all of the interrelationships that exist within nature. It would seem only prudent that human-created species should be introduced with the greatest care, and that the territory on which genetically modified plant varieties are cropped should be expanded beyond very limited levels only with great care and over extended periods of time (Leiss and Tyshenko 2002).

Recent studies have suggested that we humans now directly or indirectly (via domesticated animals) appropriate and consume a large (and ever rising) proportion of net primary production (NPP). NPP is "the energy or biomass content of plant material that has accumulated in an ecosystem over a period of time through photosynthesis. It is the amount of energy left after subtracting the respiration of primary producers (mostly plants) from the total amount of energy (mostly solar) that is fixed biologically" (Wackernagel and Rees 1996, 159). The fossil fuels that have sustained industrial society are a legacy of long-past NPP that is now slowly waning, but we already utilize the predominant share of the product of *contemporary* photosynthesis. If we expect NPP to replace the fuels that have allowed us to create the wealth of industrial society and to sustain the present human population, there will be an enormous temptation to find ways to dramatically increase NPP, to transform nature in qualitatively new ways.

The risks are likely considerable but so are the risks of not proceeding. Proceeding with molecular genetic manipulation of the plants and animals that provide our food and fibre changes the parameters and the very meaning of sustainability. Genetic modification (of either microbial species or plant species) could, however, open relatively sustainable possibilities regarding the production of transport fuels. They could also reduce the need to broadcast chemical pesticides and herbicides. Perhaps the diminution of other environmental risks should be a requirement for approving genetically modified species (though not, of course, the only one). More important, especially for Canada, we must ask what becomes of the ecosystems that might otherwise exist on the vast tracts of land that fuel alcohol wood crops might occupy, in terms of both the products that they presently produce and the nature that they sustain.

Sustainable Production, Measurement, and Policy Instruments

What should be plain from the foregoing reviews is that a sustainable economy involves all industrial sectors and virtually all dimensions of our economy and society. It is not selected economic sectors, products, or processes that must engage in sustainable production practices (or avoid non-sustainable activities); it is the whole of the economy, and both production and consumption, that must make some adjustment. There is, as well, no quick fix – the process of change must go on into the indefinite future. For me, this reality suggests a great deal about instrument choice and the role of government. Even those who accept the need for a strong role for government acknowledge that governments are not adept at micromanaging whole economies. Detailed sustainability rules and regulations are, in general, inappropriate and likely to be ineffective. The primary reason for this is that no one or several regulatory interventions would significantly alter sustainability outcomes. Achieving a sustainable economy would require many thousands of such interventions.

One exception to this general doubt about regulatory intervention might be a fleet average fuel efficiency standard for new vehicles that included all cars, vans, small trucks, and SUVs that required an overall improving standard of efficiency. As is well known, North American passenger car fuel efficiency has improved over the past several decades in conformity with the so-called Corporate Average Fuel Economy (CAFE) standards (by 33 percent between 1970 and 1995) (MacKenzie 1997, 126). Trucks, however, have shown a 30-50 percent increase in fuel consumption per vehicle over the same period, and have sharply increased in sales, in part because SUVs and other such vehicles have been classified as trucks and thereby exempted from fuel efficiency standards. Incidentally, overall motor vehicle fuel use in the US has increased by 50 percent over this same period, primarily because the size of the fleet has increased by 85 percent (by 44 percent on a per capita basis) (MacKenzie 1997, 127). Where it is politically possible, regulation in this area makes sense because of the enormous contribution to sustainability this one regulation might make, and the fact that in North

America vehicles contribute 25 percent of all carbon emissions. Otherwise, government's role is crucial but regulation is not necessarily the instrument of choice (although California may soon attempt it). In a seeming contradiction, the automobile industry extravagantly resists such initiatives at every turn while at the same time investing hundreds of millions of dollars in more sustainable alternatives to the internal combustion engine.

More broadly, government's first task is to help to establish internationally accepted ways to effectively define and measure sustainability, including materials and energy throughputs. Measurements should be standardized so that performance can be compared sector by sector, across borders, and through time. We need to know whether our transportation system, for example, is more or less sustainable than that of other nations, and why. Significant initiatives have already been taken in this regard by some public and non-governmental organizations (NGOs). Canada has not been involved in the cooperative initiative involving NGO and/or governmental agencies from the Netherlands, Japan, the US, Austria, and Germany, although other measurement efforts are being undertaken by Statistics Canada and the International Institute for Sustainable Development (IISD) in Winnipeg (Matthews et al. 2000 is an example of the former initiative). The challenge is to make such measurements as important as other economic factors in producer, consumer, and governmental decisions – to be understood, in effect, as "non-dollar economics."

Sustainable production is a highly complex undertaking and must be achieved on a product-by-product basis, on a sector-by-sector basis, and within an economy as a whole through shifts in the relative weighting of sectors. On a product or single-process basis, the best measurement technique is life cycle analysis (LCA) (Paehlke 1999). This method has been widely used in industry and has served as a partial basis for decision making regarding beverage containers and a host of other products. In Europe, LCAs provide the basis for obtaining "green product" status. Whole economies (of cities or nations) or particular economic sectors can be compared in terms of ecological footprint analysis or total material requirements (TMR). Ecological footprints are more visual and are therefore especially useful in public education. TMR measurements, however, are more straightforward and may be more likely as a standard economic policy measure.

Thus far, TMR treats all throughputs in an undifferentiated manner. Analysts, however, are now increasingly inclined to distinguish among several levels of environmental impacts associated with particular classes of materials and energy sources. One possibility that might be suggested here would see a continuing use of an overall TMR measure, supplemented by: (1) a measurement of materials problematic in sustainability terms (oil, natural gas, fish, high-grade wood), and (2) a measurement of products whose extraction, production, or use may involve a high level of ecological or toxicity impacts (including human health effects such as those associated with asbestos or metal smelting, chemical toxicity, or special ecological problems).

Measurement alone will not necessarily lead to sustainable production innovations, but it can help us understand the meaning of sustainable economies. Measurement also provides a better sense of which designs and changed habits produce significant sustainability gains and which do not. Such enhanced understanding is in turn necessary to the development of policy tools designed to accelerate sustainable production initiatives. I am unorthodox enough to believe that governments, acting collectively in a global age, may need to "macromanage" sustainability performance. While micromanagement is both unnecessary and well beyond the political pale, it does not follow that markets alone are adequate to the task at hand, if for no other reason than that market (and for that matter electoral) horizons are far too short. Thinking globally and in the long term is called leadership, and there is no good reason why Canada could not play such a role, as it once did.

The best "negative" policy option, in my view, is negotiated price levels for selected resources, or even the establishment of commodity prices pegged to that of manufactured goods. The best "positive" options are incentive programs for performance improvements in selected industries (the application of the refrigerator design initiative to other products and sectors). Either of these options would leave all of the details of product decisions and choices to industry and consumers. There is no pretence, however, that long-entrenched behaviours will somehow magically change without sustained economic (based on price and reward) motivation. Such incentives and disincentives are highly effective, especially with greater forewarning regarding the future than markets alone are likely to signal. We know enough already to act collectively in an anticipatory manner in some cases (energy, fossil fuels, and old-growth timber, for example). That may be sufficient. If not, other interventions could be developed as we learn to think in three dimensions rather than one. We should recognize that doing so may not make us richer in the short run, but could advance our well-being as a species in the long run.

Sustainable Development and Competitiveness

One might observe that no society is likely to go broke in the long run overestimating the importance of sustainable production innovations. The hedge and the uncertainty in this are, of course, on the time frame. It is not clear how long it will take for the rest of the world to catch up with sustainable production innovators. That is, the near-term level of demand for, and the price such innovations will command, is not predictable without the pricing interventions of the sort suggested above. I find it hard to imagine a positive medium-term (twenty-to-thirty-year) future wherein the price of energy and some materials will not increase disproportionately. That is, we cannot be sure or anything like sure that post-oil energy sources will be either cheaper or cleaner than oil and natural gas. Until there is some tangible evidence to the contrary, we should do our best to accelerate the complex transition over an extended period rather than wait until markets or military decision makers force extreme environmental and social outcomes over a shorter one.

Betting heavily on sustainable production may even also be our best economic option in many cases. In the early 1990s, as chair of the Peterborough (Mayor's) Committee on Sustainable Development (PCSD), I had the opportunity to see some of the competitive advantages inherent in sustainability innovations. One of the most striking was visible on a tour of an industrial-scale laundry in Peterborough (which did the washing of hospital linens for our region). This industry had been nominated for a PCSD award in part because it had recently invested a million dollars on a massive new washing machine that cleaned and reused water and recaptured heat (from very hot water, given what they were washing). The savings in both energy and water were enormous. As well, they lowered their overall costs sufficiently to remain competitive with throwaway items for many hospital functions (another sustainability gain). Thus, sustainability was enhanced in at least three ways. The economic bottom line, however, was that this very expensive washing machine the size of a large ranch house was imported from Germany.

Afterwards, I realized that given higher average European energy prices, it only made sense that this sophisticated piece of capital equipment would be developed there in the first instance. Once designed and built, each additional machine could be produced for far less than it could be developed and built in North America, where we had systematically kept energy prices lower to *advantage* our domestic industries. One can only conclude that nations and industries that anticipate the resource limitations of tomorrow are likely to prosper in the face of global adversity. Nations like Canada, with a vast land mass relative to population, may disproportionately benefit from a world more oriented toward sustainable development, both because of the need to turn to renewable resources and because we will suffer less imported pollution. Even if Canada has a relative abundance of water, trees, and energy, it is in our best interest – environmentally, socially, and, in the long run, economically – to proceed as if we did not.

Note

1 Including Quebec City, Montreal, Toronto, Kitchener-Waterloo, Ottawa, Vancouver, Victoria, and others.

References

Boyd, D. 2001. What's in the fridge? *Globe and Mail,* 18 January, A19.

Carley, M., and P. Spapens. 1998. *Sharing the world.* London: Earthscan.

Colburn, T., D. Dumanoski, and J. Peterson Myers. 1996. *Our stolen future.* New York: Penguin.

Cousins, E., R. Perks, and W. Warren. 2005. *Rewriting the rules: The Bush administration's first-term environmental record.* Washington, DC: Natural Resources Defense Council.

Dower, R., D. Ditz, P. Faeth, N. Johnson, K. Kozloff, and J.J. MacKenzie, eds. 1997. *Frontiers of sustainability.* Washington, DC: Island Press.

Durning, A., and Y. Bauman. 1998. *Tax shift.* Seattle: Northwest Environment Watch.

Faeth, P. 1997. Sustainability and US agriculture. In *Frontiers of sustainability,* ed. R. Dower et al., 47-120. Washington, DC: Island Press.

Fischer-Kowalski, M., and H. Haberl. 1998. Sustainable development: Socio-economic metabolism and colonization of nature. *International Social Science Journal* 158: 573-87.

Johnson, N., and D. Ditz. 1997. Challenges to sustainability in the forest sector. In *Frontiers of sustainability,* ed. R. Dower et al., 191-280. Washington, DC: Island Press.

Leiss, W., and M. Tyshenko. 2002. Some aspects of the "new biotechnology" and its regulation in Canada. In *Canadian Environmental Policy,* ed. D. VanNijnatten and R. Boardman, 321-43. Toronto: Oxford University Press.

Lovins, A., et al. 1979. Special issue on soft energy paths. *Alternatives* 8 (Summer/Fall): 4-9.

Lugar, R., and J. Woolsey. 1999. The new petroleum. *Foreign Affairs* 78 (January/February): 88-102.

MacKenzie, J. 1997. Driving the road to sustainable ground transportation. In *Frontiers of sustainability,* ed. R. Dower et al., 121-90. Washington, DC: Island Press.

McGuinty, D. 2000. Sustainable development in the new millennium. *Horizons* 3 (November): 3-4.

Matthews, E., et al. 2000. *The weight of nations.* Washington, DC: World Resources Institute.

Newman, P., and J. Kenworthy. 1999. *Sustainability and cities: Overcoming automobile dependence.* Washington, DC: Island Press.

New York Times. 2005. Environmentalists protest Apple's *i*waste. *New York Times,* 12 January.

Organisation for Economic Co-operation and Development (OECD). 2000. *Economic surveys: Canada.* Paris: OECD.

Paehlke, R. 1991. *The environmental effects of urban intensification.* Toronto: Ministry of Municipal Affairs.

–. 1994. *Building materials in the context of sustainable development: Ecological impacts of resource extraction.* Ottawa: Forintek Canada.

–. 1999. Towards defining, measuring and achieving sustainability: Tools and strategies for environmental valuation. In *Sustainability and the social sciences,* ed. E. Becker and T. Jahn, 243-64. London: Zed Books.

–. 2003. *Democracy's dilemma: Environment, social equity and the global economy.* Cambridge, MA: MIT Press.

Portnoy, K. 2003. *Taking sustainable cities seriously.* Cambridge, MA: MIT Press.

Robinson, J., and J. Tinker. 1997. Reconciling ecological, economic and social imperatives: A new conceptual framework. In *Surviving globalism: The social and environmental challenges,* ed. T. Schrecker, 71-94. London: Macmillan.

Roodman, D. 1996. *Paying the piper: Subsidies, politics, and the environment.* Washington, DC: Worldwatch Institute.

Ross, R. 2005. Toxic exportation. *Toronto Star,* 3 January, D1.

Scientific American. 1998. Special report: The end of cheap oil. March, 77-95.

von Weizsäcker, E., A. Lovins, and H. Lovins. 1998. *Factor four: Doubling wealth, halving resource use.* London: Earthscan.

Wackernagel, M., and W. Rees. 1996. *Our ecological footprint: Reducing human impact on Earth.* Gabriola Island, BC: New Society Publishers.

Wilson, J. 1998. *Talk and log: Wilderness politics in British Columbia.* Vancouver: UBC Press.

Part 2
The Knowledge-Based Economy, Social Capital, and Product Design

4
Developing Sustainability in the Knowledge-Based Economy: Prospects and Potential

Keith Newton and John Besley

The issue of sustainable development has acquired increasing prominence on the public (and, as will be shown, private) policy agenda in recent years. It demands responses from all players in the socioeconomic system. This chapter focuses on responses by firms[1] in the form of various combinations of approaches to sustainable production (SP). A major objective is to place the discussion of SP squarely in the context of the unfolding techno-economic paradigm (TEP) known as the knowledge-based economy (KBE). It will be argued that this new TEP brings with it unprecedented environmental concerns and challenges stemming from the unprecedented volumes of production and consumption of an ever-increasing population. At the same time, the KBE holds out the prospect of being able to address the planetary ecological threats through growing awareness, knowledge, political will, scientific and technological advance, and the resources to marshal them.

More specifically, this chapter contends that certain features of the KBE pose particular challenges to firms. We suggest that these characteristics and challenges of the KBE logically call for certain strategic corporate responses in the form of a set of management practices that constitute a new business paradigm. The drive for sustainable production is shown to be a logical and important component of that set. Accordingly, the chapter is organized as follows. The first section describes some of the major trends, forces, and features of the evolving global KBE, emphasizing those of particular relevance to the concept of sustainable production. The second section looks more closely at the KBE and sustainability. After briefly setting out some of the challenges, it develops a model of SP as an essential component of a new business paradigm that is a logical response to the exigencies of the KBE. It then examines the prospects and potential for successfully pursuing SP in terms of strategies, indicators, resources, and tools. The prospects and potential theme is then further considered in terms of the contribution of information and communication technologies (ICTs) to the pursuit of SP. Finally, in the fourth section, a simple economic model is developed to examine trade-offs and complementarities between quantitative (largely pecuniary) and qualitative (including environmental) societal objectives.

Trends, Forces, and Features of the KBE

The knowledge-based economy is widely acknowledged to be no less than a new techno-economic paradigm that is revolutionary in scale, impact, and pervasiveness. An essential characteristic is the central role of knowledge (tacit and codified) in the cumulative process of innovation. The increasing knowledge-intensity of economic activity reflects this (Gera et al. 2001; OECD 1997b). Certain features of the evolving KBE are of particular relevance to the issue of sustainability. First, it is global in scope. Trade liberalization has led to growth rates of world trade in recent decades that far surpass the growth of world GDP, and the growth of foreign direct investment has been even greater. Global corporations undertake economic activities around the globe and inter-firm alliances and networks have burgeoned and are now the typical operating mode in the KBE. Trillions of dollars of financial transactions are undertaken daily.

A second major driving force is rapid, incessant, and pervasive technological change, exemplified in particular by the advances in biotechnology and ICTs. The latter constitute a powerful revolutionary force: they are "general purpose," "enabling" technologies that are ubiquitous and have myriad applications. Their speed and computing power provide researchers with enormous analytical potential and the ability to rapidly exchange new knowledge and ideas. As networks, they bring information to millions of people around the world and play an important role in heightening awareness of issues and shaping public opinion.

The KBE and Sustainability

The Challenges

The scope and magnitude of the environmental problems besetting global society are well known and will not be treated at length here. Suffice it to say that concerns surround the observation that economic development has been accompanied by problems ranging from water contamination, soil erosion, climate change, and ozone depletion to the accumulation of persistent organic pollutants.

At the same time, certain features of the KBE – the emphasis on scientific and technological advance, continuous innovation, increased educational attainment, growing public concerns, the analytical and communication power of ICTs, dematerialization, and growing understanding of the mutually complementary objectives of eco-efficiency in firm performance – hold out the prospect of successfully addressing the admittedly daunting challenges. Our focus is on the last of these features – namely, sustainable production and its variants and guises, including eco-efficiency.

Toward a Model of Sustainable Production

This section develops a contingency theoretical model in which a nexus of contextual characteristics of the KBE elicit a set of responses by firms in the form of a variety of innovative strategies, processes, and practices. Strategies relating to sustainable production are shown to be mutually complementary to a variety of other innovative management practices in achieving salutary effects on a variety of performance outcomes.

The basic thesis advanced in this section is that a powerful confluence of forces evolving in the KBE demands a certain set of responses by various players – individuals, firms, and institutions of various kinds, including governments. As our principal focus is on sustainable production, we concentrate on responses at the level of the firm. The model depicted in Figure 4.1 suggests that a set of contextual forces in the KBE are manifested in certain challenging characteristics that, in turn, call for the acquisition of certain skills and attitudes on the part of the managers of firms. Thus, for example, flexibility is required for quick responses to emerging opportunities. Opportunism is enhanced by constant scanning and networking, which must increasingly transcend national boundaries to reflect a "global mindset." Once perceived, opportunities and challenges must be met with

Figure 4.1

Evolution of the new paradigm: responses to the challenges of the knowledge-based economy

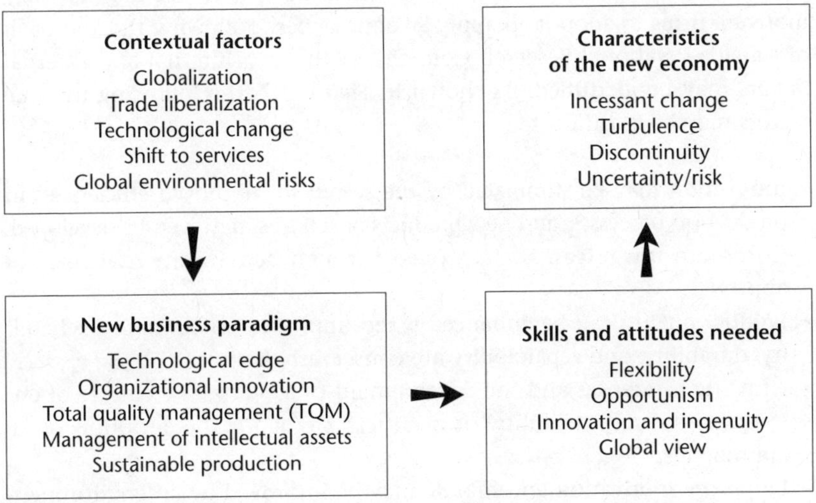

ingenuity,[2] creativity, and innovation. The acquisition and development of these skills, attitudes, and instincts need, in turn, to be facilitated by a framework or strategy, which we refer to as a "new business paradigm."

This paradigm, as can be seen in the upper right quadrant of the figure, has several components. A winning business strategy in the innovation-based KBE places a premium on the quest for, and maintenance of, technological edge, for example. Second, complementary, mutually reinforcing organizational innovations are needed to realize the potential of the "hard" technology. Third, given the rigours of highly discriminatory global markets that demand customization, uniqueness, and variety, some variant of total quality management (TQM) is required to satisfy the market. Next, knowledge production, management, and leverage are the keys to innovation in the KBE. Knowledge workers are an increasingly important factor of production. Intellectual and other intangible assets (for example, "image assets" such as good corporate citizenship and "greenness") are increasingly valued explicitly by the market, as recent estimates of Tobin's *q* attest.[3] Finally, as we will develop below, there is growing recognition that in the urgent and incessant quest for new efficiencies, innovation, and competitive edge, various elements of sustainable production have salutary effects on a firm's performance and are complementary to other components of the new business strategy. Technological edge, organizational innovation, and TQM, for example, are clearly complementary to a sustainable production emphasis, but so too are some elements of the concepts of organizational learning and management of intellectual assets.

We have alluded somewhat obliquely to the fact that there is accumulating empirical evidence of the positive performance impacts of sustainable production. What, then, are some of the more specific factors that may motivate firms to adopt and apply SP approaches? Following the approach of Canada's National Research Council (2000), both internal and external factors may be identified. As shown in Figure 4.2, the following internal factors may be significant:

- Innovation may be stimulated by the search for improved efficiencies in production processes, and new products or services may even be developed.
- Corporate image can be improved through "environmental seals of approval."
- Product quality can be enhanced when improved functionality, reliability, durability, and repairability are consciously built in.
- Growing awareness and the development of a "corporate culture of environmental responsibility" is a critical factor for the adoption of SP approaches.
- Employee motivation and morale may be improved by an environmentally responsible corporate culture.

Figure 4.2

Internal factors motivating firms to adopt sustainable production approaches

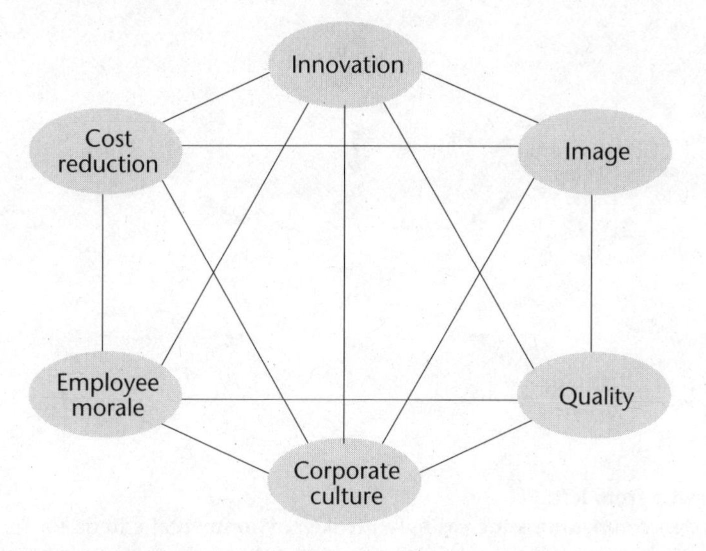

Source: Adapted by the authors from the National Research Council Canada's DfE tool for small and medium-sized enterprises (SMEs) under the NRC's Industrial Research Assistance Program (IRAP).

- Cost reduction can be effected as fewer and smaller material and energy inputs are used and waste is reduced.

The interdependencies among these factors are suggested by the lattice of linkages in the diagram. Thus, for example, an environmentally responsible corporate culture is good for employee morale; it expresses concern for corporate image, pursues quality, and encourages innovation. Innovations can enhance a company's reputation with clients and employers, while also improving quality, lowering costs, and reinforcing corporate culture. Clearly, corporate leadership (and the dearth and poverty thereof) is a critical issue in this regard.

Several external factors may also be at work, as depicted in Figure 4.3. In this section we placed the concept of sustainable production in the context of the characteristics and challenges of the fiercely competitive, innovation-driven global KBE. We have shown that SP is a logical component of a new business paradigm and is complementary to other important elements in that strategic framework, and that there are solid performance-based reasons for its adoption. To round out this model, we move to a lower level of generalization and suggest that, in operational terms, implementation of

Figure 4.3

External factors motivating firms to adopt sustainable production practices

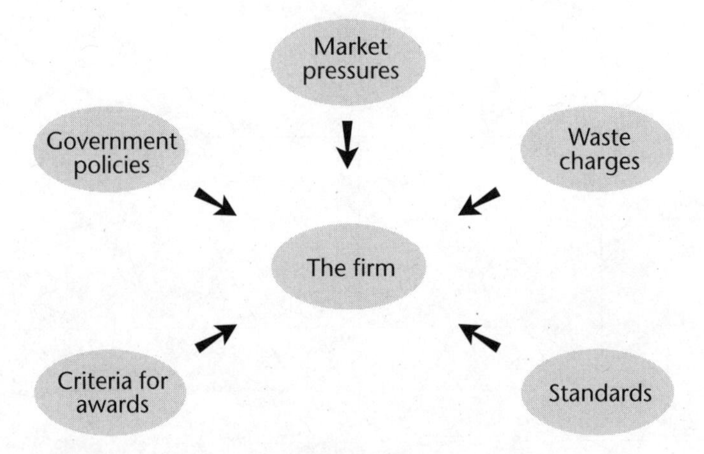

Clockwise from left:

- Design competitions increasingly invoke environmental criteria for their awards.
- Government policies of various kinds may be a powerful motivator. Examples include: "take-back" requirements for major consumer durables; eco-labelling programs; requirements for provision of environmental information; subsidies for research on environmental technologies.
- Market pressures may come in various forms: large firms may impose standards on their suppliers; growing consumer preference for "green" products and services, and/or direct pressures such as product boycotts; growing demand for the use of environmental criteria in consumer reports.
- Waste disposal and processing charges are becoming significant enough to encourage firms to pursue practices of waste reduction, reuse, and recycling.
- Trade associations may pressure members to adopt certain norms (e.g., the Responsible Care program of the Canadian Chemical Producers Association), and standards organizations may be influential (as in the case of the introduction of the ISO 14000 series).

Source: Adapted by the authors from the NRC's DfE tool for SMEs under IRAP.

SP requires systematic analysis of the firm's product life cycle(s). The latter typically has five phases, each of which may be examined for its potential contribution to SP. As depicted in Figure 4.4, the stages are design, production, distribution, product use, and end of cycle.

The critical role of design is immediately apparent – design in two senses. First, the product design itself will determine to some extent the available set of possibilities for technologies, materials, and so on. The product design

Figure 4.4

The product life cycle

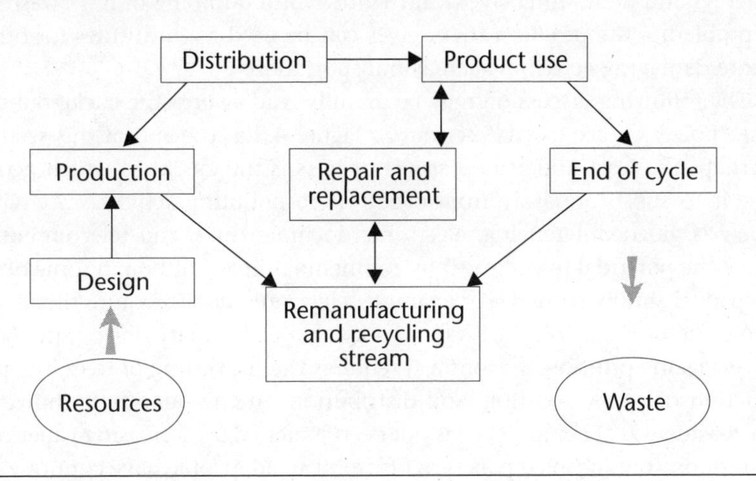

Source: NRC 2000.

will also determine the features of functionality, reliability, durability, and repairability, all of which are critical to the product use stage. Second, process design is important since there is often "more than one way to skin a cat,"[4] so choice of process(es) will also be important in determining the firm's eco-efficiency. Little wonder, then, that a principal component of sustainable production strategy is Design for Environment (DfE) (see Chapter 6).

The Role of ICTs in Promoting Sustainable Production

Other chapters in this book examine some of the institutional infrastructure associated with the emergence of the concepts of eco-efficiency, eco-effectiveness, life cycle analysis (LCA), total quality environmental management (TQEM), ISO 14000, DfE, and so on. They look at models, tools, indicators, resources (such as inventories of environmental technologies, best practices, case studies, and so on) and institutions that may potentially promote the uptake and diffusion of SP strategies. Many of these resources are seen to reside in international institutions. The challenge of building Canadian capacity will depend in part on our ability to use, refine, and extend them for use in specifically Canadian situations. As an extension of the "potential/capacity" theme of this volume, we look at the ability of ICTs to contribute to this process.

On the question of the potential role of ICTs in promoting sustainable production, it is clear that such technologies can play a powerful role in the associated networking and information dissemination. ICTs have improved the ability of ecological researchers to conduct the monitoring and evaluation necessary to explain the increasing urgency of seeking sustainability. ICTs may be a mixed blessing, however. While manufacturing of microelectronics is energy- and toxics-intensive – and Third World dumping of ICT "waste" is a problem – the products themselves can be used as substitutes for other, more damaging activities such as transportation.

The following discussion may be usefully read against the background of the "policy choice matrix" set out in Figure 4.5 at the end of this section. Perhaps the least understood aspect of ICTs is the extent to which people use it to substitute away from other, more polluting activities. Activities such as videoconferencing, electronic document use, and telecommuting have the potential to decrease environmental impact, although some of the impact is simply shifted. For example, electronic books require the manufacturing of electronic hardware and consume electricity during use. Book or magazine printing, by contrast, requires the harvesting of trees, the production of paper, printing, and distribution. At the same time, however, increasing use of computers has not been matched by decreasing paper consumption (the so-called paperless office). Instead, people have continued to prefer reading most texts off paper. Abramovitz and Mattoon (2000) describe how paper use is correlated with wealth, and cite industry groups to say that paper consumption will likely rise by 31 percent by 2010 after having already doubled six times since the 1950s. The authors point to PaperCom (http://www.papercom.org) as a US industry group that has actively tracked the relationship between paper and the adoption of Internet technologies, concluding that the electronic era will create new markets and opportunities for many, if not most, paper-based products and services. And while computers and the Internet are emerging as important media, the measures we have today strongly suggest that paper will continue as a core industry for decades to come.

Handheld technologies or new technologies that make computer screens easier to read continue to promise an end to the near monopoly of the printed page. The much-touted potential of videoconferencing seems equally troubled as passenger airline travel continues to climb. While information technology has made daily face-to-face contact unnecessary, the increasingly global, integrated, and ever-wealthier societies have still led to increasing levels of travel, both by air and by road. In addition to business travel, tourism has emerged as a growth sector of the world economy. To the affluent denizens of the KBE, books and video of foreign places no longer suffice. Global airlines and western-quality amenities means that even the elderly can visit the far corners of the globe without fear for their health or safety. The difference between the current situation and what existed in the not-so-distant past is the range of choices available to those who inform themselves of their options. ICTs make discovery of those choices increasingly simple. These choices – between travelling to a conference or participating online, for example – will largely determine an individual's or group's environmental impact. Relative costs will play a part in this choice, but preferences or even tradition will also continue to factor into decisions.

A less equivocal area where ICTs can be used to support sustainability is that of monitoring and assessment. The United Nations Environment Programme manages a core program on Environmental Assessment, Information and Early Warning. Central to this program is work being done through its Global Resource Information Database (GRID). The GRID system includes a number of nodes that create and publish state-of-the-environment (SoE) reports and rely extensively on Geographic Information Systems (GIS). When combined with the increasing array of military and non-military satellites that beam down data from space, GIS represents a potent force for understanding the larger workings of environments. Although Canada does not have a GRID node, Natural Resources Canada operates the Canadian Centre for Remote Sensing. The centre uses the Canadian Space Agency's RADARSAT technology as well as satellite data from around the world to monitor environmental issues.

Another use of ICTs that has changed the context within which environmental monitoring is done is in public access to information on toxics. Several countries have developed what the OECD has called the Pollutant Release and Transfer Registry, which complied with a set of guidelines developed by the organization and published in 1996. In Canada, the National Pollutant Release Inventory (NPRI) requires emitters of listed substances to report on their emissions. These are then made public by the government (http://www.ec.gc.ca/pdb/npri), enabling residents to search for polluters near them (see Chapter 7).

One of the most important aspects of ICTs in promoting sustainable development is the role that networked information systems can play in providing a forum where interested parties can obtain, discuss, and even create new knowledge. The Metcalfe Principle, named after the founder of Ethernet developer 3Com, hints at the importance of networks in creating a knowledge-based society. It reads: "The value of a network grows as the square of the number of its users. Or, more plainly stated: The more users who can talk to each other on a network, the more valuable it is" (http://www.wired.com).

In the advocacy field, the increasing value of networking was demonstrated most recently in the organizational process that took place in the lead-up to the World Trade Organization's 1999 meeting in Seattle. ICTs helped a coalition of local, national, and international organizations make an impact on the WTO and the mindset of the conference's participants. These organizers used the Internet to both provide information and coordinate real-world events. In many ways, this chapter itself is a product of the power of the Internet to provide environmental information. Were it written as recently as eight years ago, much of the information cited would not have been as readily available, particularly the information from governments and the media. If individuals need to be aware of their options so that they can choose the environmentally responsible course of action, then the Internet is perhaps the natural world's greatest ally.

Protesters are not the only ones who can use the power of networks to create change. Beginning in late 1995, Maurice Strong chaired a task force brought together by the International Development Research Centre (IDRC), the International Institute for Sustainable Development, and the North-South Institute to discuss "Canadian strengths and capabilities with regard to the global development challenges" of the new century. The report he tabled in November 1996 concluded that knowledge sharing and knowledge development would be the key to prosperity in the twenty-first century, for developing and developed countries alike. As such, the task force recommended that Canada establish a role for itself as an international "knowledge broker" and draw on its strength in the telecommunications field to help build capacity within targeted countries to take advantage of this knowledge: "The task force recommends, as a matter of urgency, that *knowledge,* and the communication and information technologies that can advance knowledge, be placed front and centre in Canadian foreign policy and Canada's international outreach. Canada should position itself for the coming century as a creator and as a broker of knowledge for sustainable development" (IDRC 1996).

Besley (2000) describes in greater detail some of the many powerful electronic networks and databases now emerging on sustainability issues. He observes that Google, a powerful Internet search engine, returns more than 500,000 site hits for the search term "sustainable development," and comments that, if anything, the sheer volume of available information may have the potential to drown those who choose to plug in. The potential of information overload – not just in the environmental field – represents an important outcome of the Internet age and a favourite topic of the business media. Nevertheless, "above all, the information technologies disrupt hierarchies, spreading power among more people and groups. In drastically lowering the costs of communication, consultation, and co-ordination, they favour decentralized networks over other modes of organization. In a network, individuals or groups link for joint action without building a physical or formal institutional presence."

In Canada, former environment minister David Anderson (2000) noted that young people will not be content simply to wait for the conventional political process to make slow progress on these issues. Thanks to communications technology, a new kind of e-democracy is emerging, driven by the young, that public decision makers will be unable to ignore in the future.

Figure 4.5

What will the KBE do? A policy choice matrix

The KBE may advance sustainability by:	The KBE may impede sustainability by:	Policy challenges
Helping society understand innovation and create products and processes that responsibly improve quality of life	Allowing society to innovate in environmentally harmful ways	Choose instruments that will be used to ensure that innovations are made in areas that lead to sustainability.
Helping society track and respond to indicators of sustainable production and consumption	Encouraging increased resource use in production and consumption	Develop and employ instruments that will ensure that capital is used in a sustainable way.
Helping to diffuse sustainable technologies	Helping to diffuse non-sustainable technologies	Utilize instruments that discourage the use of non-sustainable technologies.
Emphasizing the role of education and lifelong learning	Emphasizing learning only as a tool for wealth creation	Ensure that principles of sustainability will be incorporated into all levels and types of education.
Helping poorer countries develop beyond the subsistence level	Enabling ever-greater inequities in wealth (e.g., a "winner-takes-all" society)	How can we achieve an equitable distribution of wealth?
Emphasizing the importance of developing effective cities	Pushing for quick growth that ignores non-economic aspects of community	How do we want to live, and what tools can help us achieve that goal?
Encouraging trade based on a sustainability concept of competitive advantage and not on perverse subsidies	Enabling trade that is not based on principles of sustainability	What national and international rules do we want to create and enforce to ensure that sustainability is a more explicit criterion in international trade relations?
Helping societies understand their natural environments	Allowing societies to find and exploit remaining stockpiles of renewable and non-renewable resources	What knowledge about sustainability will be critical and how can it best be diffused?

Conclusion:
Tilting the Big Trade-offs in Favour of Environmental Goals

It may be useful to think of the role of innovation policy in addressing the trade-offs and complementarities among broad societal goals – what the OECD, quoted above, has called *quantitative and qualitative* socioeconomic goals. Once again, the context is the growing wealth potential, consumption, production, and technological prowess of the KBE. The policy dilemma is: how do we use this enormous potential, to what end, and with what results?

As a starting point for a discussion of policy considerations, we posit a simple model relating to two sets of possible societal objectives. Thus, in Figure 4.6A, it is suggested that the economic resources of the society might be used to produce two broad categories of "goods."[5] The first category we call SBO (for sheer bloomin' opulence); it consists of all the accoutrements of wealth: SUVs, snowmobiles, personal watercraft, eco-tours, and other manifestations of the ability to engage in conspicuous consumption. The second is dubbed QOL for quality-of-life goods, such as health, education, the environment, culture, social capital, and so on. As depicted in Figure 4.6A, the economy is capable of producing the various combinations of these "goods" in accordance with its production possibilities frontier (PPF). The PPF thus indicates, in a stylized manner, the trade-offs among the combinations of goals of which the economy is *capable,* given its finite resources. One may next posit the notion of an "aggregate indifference curve" or "social preference function" (SPF) showing society's trade-offs among *preferred* combinations of the goods. The superimposition of the PPF and the SPF as in Figure 4.6B gives a tangency point reconciling the economy's capacity to produce the two goods and society's preferences for them.

Now consider the role of innovation and of policy, respectively. The important point about innovation, of course, is that it is the key to improvements in efficiency, productivity, and competitiveness. In terms of our model, therefore, it is a means to expand productive potential: in other words to "push out" the PPF as in the "parallel shift" in Figure 4.6A – from PPF_1 to PPF_2. It is then immediately apparent that the available choice set for society is thereby enlarged – citizens can potentially enjoy greater amounts of SBO and QOL simultaneously.

What, then, is the role for policy? In this simple model, the role is potentially twofold. First, given political will exercised (one supposes) through a democratic process, public education, suasion, and advocacy could be used[6] to heighten societal consciousness and awareness about the importance of, and the relative priority to be accorded to, QOL (and especially environmental) goods. Diagrammatically, what this means is "tilting the social preference function": the society now accords greater weight to QOL goods than formerly. The trade-off (or "marginal rate of substitution") is changed,

Figure 4.6

Improving the trade-offs among societal goals: the roles of innovation and policy

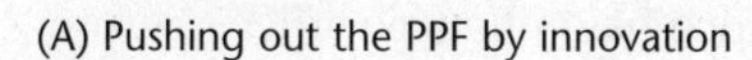
(A) Pushing out the PPF by innovation

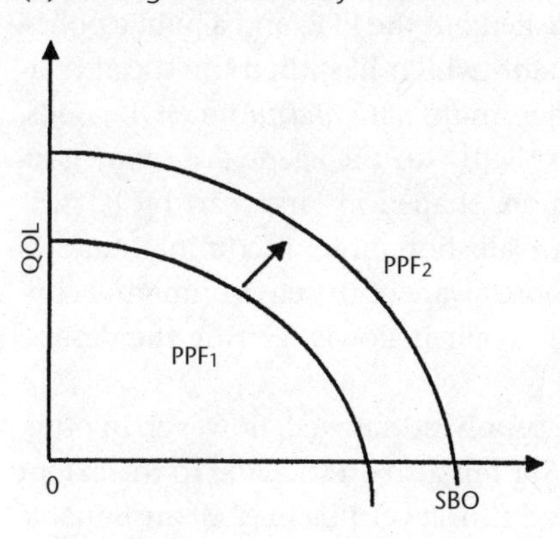

(B) Adding the SPF

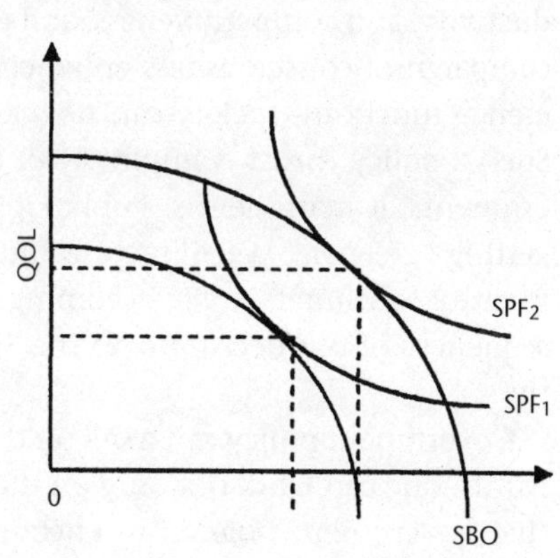

(C) Tilt 1: A shift in preferences

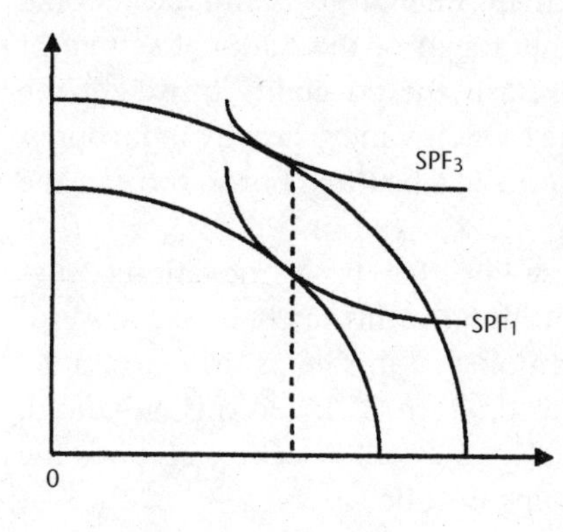

(D) Tilt 2: A shift in productive capacity

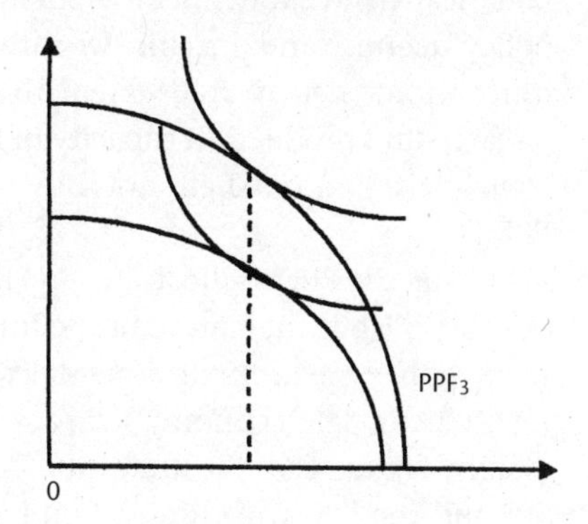

(E) A double tilt

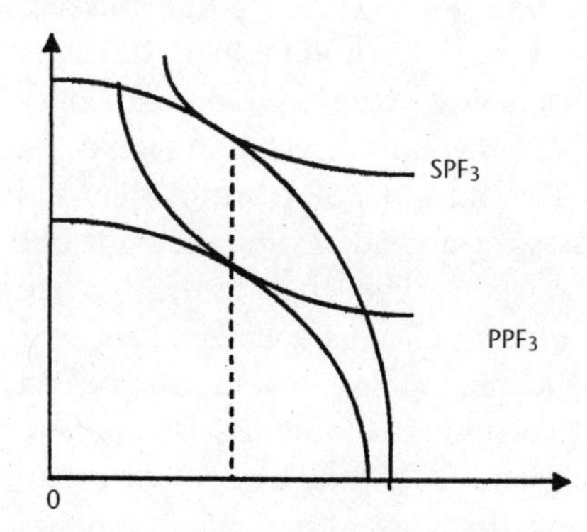

and graphically the shape of the SPF, as in Figure 4.6C, reflects this. With the same PPF configuration as in Figure 4.6B, public policy has swayed societal preferences toward QOL goals. Equilibrium now affords the same level of SBO as formerly, but with more QOL.

Second, with the help of innovation (which has enhanced efficiency, productivity, and competitiveness, and pushed out the PPF) and a public policy campaign of consciousness enhancement (which has tilted the social preference function), society can now enjoy more SBO *and* more QOL goods. Such a policy thrust is implicitly or explicitly on the agenda of many governments at many levels. Public opinion, shaped in large part by ICTs, is getting "greener." We all recycle and we all shop more discriminatingly. As wealthy consumers, we're becoming more aware of the environmental consequences of our decisions. So the SPF is, albeit slowly, getting the desired tilt.

Government policy can work on the supply side as well, however. In other words, one can work not only on the SPF but also on the PPF. To the extent that government policy can encourage firm-level practices of sustainable production (or eco-efficiency), and to the extent that government can systematically and explicitly incorporate environmental goals into the national policy agenda, and specifically into the design of the national system of innovation, then to that extent there exists the possibility to weight the research and production capacity of the economy more heavily in favour of QOL goals. In a nutshell, not only can the SPF be tilted but so too can the PPF.

"Tilting the PPF" is illustrated in Figure 4.6D. This time innovation pushes out the PPF but the nature of the innovation is deliberately biased: government policy backs environmental technologies and helps firms to implement them. The economy's capacity to produce QOL goods is enhanced. Finally, Figure 4.6E illustrates the double tilt of policy acting on both the SPF and the PPF, with innovation pushing out the latter.

What this simple model suggests is that we have, in principle, the opportunity to implement policies to tap the vast resources of the KBE and harness its powerful technologies in pursuit of an environmentally healthier balance of societal objectives. Another heuristic device as a prologue to policy considerations is a policy choice matrix of the kind depicted in Figure 4.5. Briefly, it underlines the fact that the KBE, from the SP point of view, has inherent capacities for good and ill. Assessing and addressing the trade-offs therefore necessitates identification of key policy challenges as the first step in identifying the research needed to inform judicious policy making.

In closing, we emphasize the "K" in the KBE to make two points. The first is that rapid technological advance is the hallmark of the KBE, the route to innovation, improved productivity, competitiveness, and higher living standards. From governments' point of view, the relevant analytic and policy framework appears to incorporate the concepts of national and regional systems of innovation (NSIs and RSIs) and local SIs or "clusters." At every level, therefore, it is essential to view the policies and programs directed to these systems through the lens of the trade-offs and complementarities outlined above. If the goal of SP is to be successfully pursued, leadership will be required to tilt or bias NSI agendas to systematically incorporate environmental criteria.

The second point has to do with the process of knowledge creation. Here, as seen above, governments have a potentially important role to play. Using the power of ICTs and assuming the role of facilitator, agent, broker, catalyst, creator, compiler, and disseminator of strategic information, governments can inform citizens, firms, and institutions about crucial environmental issues. At the same time, they can make widely accessible the tools that the various actors in the socioeconomic system can use to address sustainability concerns. Finally, governments in facilitator mode can continue to enrich and extend the public/business/university partnerships that contribute to an innovation agenda that clearly articulates sustainability objectives.

Notes

1 While "firm" or "business" typically denotes a private sector organization, many of the approaches discussed subsequently apply in principle not only to both the private goods-producing and service sectors but also to the public and NGO sectors.
2 Following Homer-Dixon (2000), one might characterize the assessment of the prospects and potential for developing SD as an assessment of humankind's ability to close the "ingenuity gap."
3 Named after Nobel Prize–winning Yale economist James Tobin. It is the ratio of market value to the replacement cost of capital. For many knowledge-intensive firms, the ratio is 10:1 or greater.
4 "Equifinality" is the social scientists' way of expressing the notion that, rather than a single uniquely determined best practice, there may be a variety of routes to a given end.
5 We enclose "goods" in quotation marks for two reasons: first, because we don't just mean commodities but rather goods as opposed to "ills"; second, because some, particularly of the SBO variety, are not unequivocally good.
6 Facilitated, of course, by the ICT tools of the KBE.

References

Abramovitz, J., and A. Mattoon. 2000. Recovering the paper landscape. In *State of the world 2000,* ed. L. Brown et al., 101-20. New York: Norton/The Worldwatch Institute.

Anderson, D. 2000. The environmental challenge in the 21st century. Presentation delivered at Globe 2000, Opening Plenary. Vancouver, 22 March.

Besley, J. 2000. Sustainable production and consumption in a knowledge-based economy: Is a knowledge economy a sustainable economy? MA project, Carleton University.

Gera, S., C. Lee Sing, and K. Newton. 2001. Trends and forces in the knowledge-based economy. In *Doing business in the knowledge-based economy,* ed. L. Lefebvre, E. Lefebvre and P. Mohnen. Boston: Kluwer Academic Publishing.

Homer-Dixon, T. 2000. *The ingenuity gap.* Toronto: Knopf.

International Development Research Centre (IDRC). 1996. *Connecting with the world: Priorities for Canadian internationalism in the 21 century (task force report).* http://www.idrc.ca/en/ev-62072-201-1-DO_TOPIC.html.

National Research Council Canada (NRC). 2000. *Design for Environment guide*. Ottawa: Government of Canada.
Organisation for Economic Co-operation and Development (OECD). 1996. *The knowledge-based economy.* http://www.oecd.org/dsti/sti/s_t/inte/prod/kbe.pdf.
–. 1997a. *Sustainable consumption and production: Clarifying the concepts.* Paris: OECD.
–. 1997b. *Technologies for cleaner production and products: Towards technological transformation for sustainable development.* Paris: OECD.
–. 2000. *Technology and environment: Towards policy and environment.* http://www.olis.oecd.org/olis/1999doc.nsf/LinkTo/DSTI-STP(99)19-FINAL.

5
Sustainability, Social Capital, and the Canadian ICT Sector

David Wheeler, Kelly Thomson, and Michael A. Perkin

In this chapter, we explore the potential for Canadian information and communication technology (ICT) companies[1] to create economic, social, and ecological value with reference to two theoretical constructs: sustainability and social capital. We take a global overview and provide commentary on private and public sector applications of digital technologies around the world. We develop a Canadian perspective on sustainability, social capital, and the ICT sector, and we report on research with the sector undertaken for the multi-stakeholder "Sustainable Canada" project led by York University and funded by the Social Sciences and Humanities Research Council. We observe that Canadian ICT companies do not see the issue of sustainability as a defining characteristic of their identities or of their business strategies. There is, however, a good understanding of the importance of practical commitments to environmental and social good practice, and there is potential for ICT companies based in Canada to enter into strategic partnerships with government and civil society organizations in order to help address a range of social and ecological issues in Canada and internationally. We suggest that pursuit of "made in Canada" initiatives in sustainable development may enhance social capital in the ICT sector, which may in turn improve the reputation and competitiveness of Canadian ICT firms internationally.

In his somewhat whimsical assessment of the way the "new economy"[2] is emerging, former US labour secretary Robert Reich (2001) observed: "It's as if all sellers of all products and services have suddenly been placed next to one another in a global bazaar in which all prices, and all information about quality, was immediately apparent to all buyers." Like Adams (1997), Handy (1998), Putnam (2000), Stiglitz (2002), and numerous other commentators on the nature of life, work, and society in the rapidly globalizing economy, Reich is unconvinced that the human values guiding globalization are wholly benevolent. Indeed, he suggests a number of public policy interventions that might ameliorate the negative social impacts of the new economy.

These socioeconomic and occasionally quasi-spiritual musings have been accompanied by a growing volume of commentary examining specific societal and ecological implications of new economy phenomena such as biotechnology, nanotechnology, e-commerce, and the use of the Internet by businesses, governments, and communities. The one clear conclusion from all of this commentary is that the advent of the new economy represents a paradigm shift – with all of the potential for positive and negative phenomena, contrasts, and paradoxes that such dramatic change entails. At the heart of the change is the digital technology revolution and the growing importance of the ICT sector (Tapscott et al. 1999; Castells 2000; Moss Kanter 2001).

Clearly, the ICT sector has the *potential* to play a central role in creating a global economy that is more ecologically sustainable and socially inclusive than the former "industrial economy" (Post 2000; Park and Roome 2002; Smith 2002).

The industry's proponents argue that their technology and services can help improve economic efficiency and social inclusion across a range of other economic sectors (European Information Technology Observatory 2002). The ICT sector can help decouple economic development and progress from resource use, thereby improving living standards and quality of life, and simultaneously reducing environmental impacts such as climate change. ICT companies can also facilitate the inclusion of a far broader spectrum of the world's population in directing and enjoying the benefits of these possibilities (Sheats 2000, 2001).

In the environmental domain, ICT companies can foster sustainable development by enabling better resource and energy use and by dematerializing business transactions. It is argued that through miniaturization, ICTs create the same or better technological options with fewer resources. And as technologies move toward wireless infrastructure, the environmental impacts of ICTs could be reduced even further (World Resources Institute et al. 2002). Because of their demands for ultra-reliable electricity, they may even help usher in a revolution in the need for and possibilities of distributed energy generation and "micro power" (Dunn 2000). It is argued that industries will adopt these technologies not only because they are more eco-efficient but also because they will be necessary for creating and maintaining competitive advantage.

For example, it is asserted that the advent of ICTs may contribute significantly to the reduction of greenhouse gas emissions. It has been estimated that the potential in-built contribution to reducing the energy intensity of the US economy as a direct result of the greater efficiencies of e-commerce may be as high as 2 percent per annum (Romm et al. 1999). Finally, the industry itself is becoming more energy-efficient. According to a report of the Global e-Sustainability Initiative (2002), signatories to the Environment

Charter of the European Telecommunications Network Operators Association (ETNO) decreased energy consumption by 21 percent and fuel consumption by 26 percent between 1997 and 1999.

In the social and economic domains, the new information and communication technologies can improve social and economic conditions by facilitating coordination and cohesion in communities; by reducing the impact of geography on access to information and services related to health, education, and therefore competitiveness in the knowledge-based new economy; and finally by enabling business, government, and other interest groups to more easily include their stakeholders in decision making, regardless of location. The new technologies are generally cheaper and more decentralized than earlier technologies, and therefore, in theory, more people have the potential to access them locally. This could have significant benefits for democracy and the engagement of civil society.

Wheeler and Elkington (2001) and Tapscott and Ticoll (2003) have described the potential benefits of real-time communication of environmental and social information by business. And the World Resources Institute and others (2002) have drawn attention to the implications for corporate accountability of technologies such as real-time video streaming of corporate abuse over the Internet. Kevin Hill and John Hughes (1998) have described the use of the Internet by radical activist groups opposed to consumerism and economic globalization. The potential for maximizing the positive social and political impact of grassroots movements is very broad, ranging from humanitarian campaigns such as the banning of land mines to the growth in influence of the Zapatistas in Mexico. According to the University of Texas (2002), "the international circulation through the internet of the struggles of the Zapatistas in Chiapas, Mexico has become one of the most successful examples of the use of computer communications by grassroots social movements." In contrast, the fears of early 1990s commentators (e.g., Reich 1992; Rifkin 1995) about the possibilities for massive social exclusion and dislocation or redundancy of entire workforces caused by the advent of the digital revolution have not materialized.

However, the adoption of new technologies does not come without risks and dilemmas, both for business and for society. Canadian commentator Don Tapscott has described the "promise and perils" of the digital revolution for business and the profound catalytic effect of information and communication technologies in helping establish and develop "business webs" – interdependent networks of companies that leverage knowledge and creativity for commercial purposes (Tapscott et al. 1999). These developments significantly increase both the strategic opportunities and the competitive pressures on business. Gerry McGovern (1999) laid out a blueprint for business in helping humanize the digital age. McGovern stressed the importance of basic principles such as customer care and ethical behaviour in business. Like Reich (2001), however, McGovern notes the failure of the emergence of information and communication technologies to create a better work/life balance for ordinary workers.

The creation of these technologies themselves consumes resources and produces waste, some of it hazardous to human health. The problems of industrial pollution in Silicon Valley have been well documented (Silicon Valley Environmental Partnership 1999; Environmental Protection Agency 2002). There is an equally important and well documented concern about the potential for creating a "digital divide," that is, fostering productivity growth in the developed world that is so dramatic that the gap between developed and less developed countries widens substantially and potentially becomes so wide that it becomes virtually impossible to bridge (International Labour Organization 2001; Smith 2002). Anti-globalization groups have flagged concerns regarding cultural imperialism and homogenization as technologies spread culture from developed (and digitally advantaged) countries to less-developed countries. Finally, some social capital theorists such as Putnam (1995, 2000) raise the concern that increased digital interaction may result in reduced face-to-face contact in local communities, eroding existing "real" social capital for less sustainable digitally based social capital.

In order to explore these contrasting possibilities in greater depth and help us develop a theoretical basis for discussing the future potential for ICTs to create ecological and social value internationally and in Canada, we will now discuss in greater detail some of the phenomena associated with ICTs and their role in society.

Cases from the Present and Visions for the Future

Two leading sustainability think tanks have released a series of reports and commentary on the potential and the limitations of digital technologies for promoting sustainability in the global economy: the World Resources Institute (WRI 2001-2005) and the UK Forum for the Future (Wilsdon 2001, 2002). The World Resources Institute continues to spearhead research and data gathering in this arena.[3] In addition, in 2002, Craig Warren Smith published his research on digital corporate citizenship, tracking how approximately sixty companies responsible for 70 percent of global market share in ICTs were addressing their societal responsibilities (Smith 2002).

On the more optimistic side of the equation, quite early in the debate about the social and environmental impacts of digital technologies, the WRI helped coordinate a special supplement for *BusinessWeek* describing the outcomes of a landmark conference on the Digital Dividend held in Seattle in October 2000. The conference was attended by 300 leaders of ICT industries and included contributions from former Hewlett-Packard (HP) chair and CEO Carly Fiorina and Microsoft's Bill Gates. The supplement, like the conference, was packed with good news stories and rousing rhetoric from leading businesspeople, academics, and political figures.

Accentuating the Positive

The positive potential for the digital industries was captured in Carly Fiorina's speech to the Digital Dividend conference: "We are now at the beginning of a second renaissance, the digital renaissance. Invention is once again the prime virtue. But this time the tools for invention can be extended to every corner of the earth." These sentiments were echoed and given an environmentally positive spin by Amazon.com founder Jeff Bezos and Stephan Schmidheiny, founder of the World Business Council for Sustainable Development (WBCSD). Schmidheiny stated: "The Internet is the best tool we have had for creating wealth and redistributing the opportunities to create it since the steam engine – wealth created not from our diminishing supplies of raw materials and natural resources but from our limitless resources of creativity, intelligence and information" (*BusinessWeek* 2000).

At that time, companies like HP (through its multi-million-dollar *World e-Inclusion* project), AT&T (with its schools *Learning Network* – an internal "joint venture" between its marketing and philanthropy departments), and Real Networks (with 5 percent of pretax profit committed to "progressive social impacts of digital technologies") appeared to be very serious about backing strategic leadership rhetoric with real implementation. While noting that the scale of some of these initiatives may be adjusted subject to business conditions, Smith (2002) referred to these big-ticket programs as "signature initiatives," describing them as transcending both conventional philanthropy *and* conventional business practice.

According to Smith, in 2002 approximately half of the world's leading ICT companies had such initiatives, which were typically multi-year, multi-million-dollar programs. The projects often served to unite employee interests and commitment with strategic corporate objectives despite the rapidly changing nature and structure of the ICT sector. At the heart of the strategic value proposition to ICT businesses is the shift from direct marketing to relationship and "affinity" marketing in order to secure customer loyalty (Hegel and Armstrong 1997; Reichheld and Schefter 2000). In the case of HP, this extends to both the developed and the developing worlds (for example, with its LINCOS project aimed at digitally empowering Little Intelligent Communities). HP was not alone in identifying the long-term opportunities associated with bridging the digital divide with the global south.

Paul et al. (2004) referred to more than a thousand ICT-enabled development initiatives in the WRI database that claimed to demonstrate the potential for digital technologies to empower communities and businesses in developing countries. The projects embrace activities as diverse as telecentres, agriculture, distance education, health care, e-government, microfinance, small and medium-sized enterprise (SME) growth, handicrafts, women's empowerment, and NGO capacity building. Twenty-nine percent of projects were in Africa, 36 percent in Asia, and 15 percent in Central and South America. Many of these examples are available as mini-cases on the WRI website (http://www.digitaldividend.org/case/case.htm) and would now be described as examples of grassroots "sustainable enterprise networks" (Wheeler et al. 2005).

For example, since 1997 Grameen Phone has provided a cellular service in rural Bangladesh through local entrepreneurs, generating revenues of $1,200 per annum per cellphone and maintaining average customer bases of seventy persons. These customers use cell phones to optimize prices for their goods by selecting the best local markets for their produce before they travel to them. In India, TARAhaat seeks to create an online bazaar for communities that embraces needs as diverse as farm technology, refrigeration, and bicycles while creating potential export outlets for artisans and handicraft workers. Viatru is a "fair trade" exchange with similar ambitions with respect to providing access to international retail markets for artisans in the developing world.

Leading US business academics and civil society commentators have developed provocative theories about the potential for marketing sustainable technologies to the world's poor based on an analysis of the unmet needs of the 4 billion people on the globe who live on less than $1,000 per annum at the "base of the pyramid" (Prahalad and Hart 2002; Prahalad and Hammond 2002). Prahalad challenged the Digital Dividend conference in Seattle thus: "Don't look at the poor and say there is no hope. Selling to the poor may be more profitable than selling to you and me. This is where the future is. Opportunities are everywhere. The [digital divide] is not about lack of opportunity: it is about lack of imagination."

In our original research for this chapter, we uncovered a number of especially interesting examples of what we might describe as cases demonstrating simultaneous economic, social, and ecological gain, or "triple-bottom-line" thinking. In some cases they represent novel applications of the digital technologies to generate economic, social, and ecological value; in others, they represent particular initiatives by ICT companies. One example of the former is GreaterGood.com, which is described in Case 1.

Another example of ICTs facilitating social, environmental and, economic gain through pure application of the technology is the potential for online filings of government-required information over the Internet. In some cases this has been further facilitated by governments and software companies helping bridge the digital divide for low-income families, and in others banks have used software for online filing as a marketing incentive. Well designed e-government initiatives can reduce resource use (paper, postage, and travel

Case 1

Cause-related e-commerce from GreaterGood.com

GreaterGood.com is based in Seattle, Washington, and is a privately held, for-profit company. It owns and operates several "click-to-donate" sites, including:

- The Hunger Site – the world's first "click-to-donate" site, where over 95 million visitors donated nearly 10,000 metric tons of free food to help feed the hungry during 2000. This site was still operational in early 2005, with nearly 50 million donations of "cups" of food claimed for 2004.
- The Rainforest Site – founded on 1 May 2000 to help protect the environment. Fifteen million donations preserved over 1,900 acres of the world's rainforest in 2000. In addition, the EcologyFund Site claimed nearly 15,000 hectares of land "saved" through donations up to February 2005.
- The Kids AIDS Site – founded on 18 September 2000 to help provide HIV care, education, and counselling to pregnant women and mothers with newborns. During 2000, nearly 6 million donations led to the provision of 827 care days. This site was superseded by a more broadly based Child Health Site, which claimed nearly 600,000 children in need helped in 2004.
- The Child Survival Site – where 3.5 million donations during 2000 provided 270,000 capsules to support global supplement programs providing vitamin A to children every four to six months through preschool age. This site was also superseded by the Child Health Site.
- The Breast Cancer Site – where 4.7 million donations facilitated 440 mammography screenings for underprivileged women who otherwise would not receive the gift of early breast cancer detection during 2000. By 2004, this figure had risen to 2,135 screenings. This site was still operational in early 2005.

These sites, together with the Animal Rescue Site (nearly 35 million bowls of food donated in 2004) and the Literacy Site (over 150,000 books donated in 2004) enable visitors to participate in worldwide efforts to address social, animal welfare, and environmental problems at no extra cost to them. Advertising sponsors pay for the donations generated by clicking on the sites' "donate" buttons. In addition, GreaterGood.com operates the world's leading cause-related shopping portal (http://www.greatergood.com), where up to 15 percent of each purchase automatically benefits the charity, K-12 school, or university scholarship fund of the shopper's choice. GreaterGood.com works with over 3,500 not-for-profit organizations, including the Humane Society, the Nature Conservancy, Save the Children, and Special Olympics; and over 100 brand-name retailers, including Amazon.com, L.L. Bean, Lands' End, Nordstrom, Dell, and OfficeMax.

Case 2

Governments going online

In March 2001, the US Internal Revenue Service (IRS) claimed that 35.4 million Americans filed their federal income tax returns over the Internet in 2000. The figure for 2001 was 40.25 million, of whom over 6 million filed online from their home computers (http://www.irs.gov). The figure expected for 2003 was in excess of 50 million. In support of this, and in an effort to help bridge the digital divide, software manufacturers and financial services companies have introduced low-cost or free options for online filing for low-income families. Intuit's *Quicken Tax Freedom* program enables families with adjusted gross incomes of under US$25,000 to file their tax returns by computer for free. Companies such as WingSpanBank.com offer their customers free online access to tax filing software as a bonus for depositing money with the bank. The IRS itself sponsors a free online filing service for low- to moderate-income families, non-English speakers, and the elderly. The benefits to the IRS and the users of the systems include faster processing and significantly lower rates of error. According to Cathilea Robinett, executive director of the California-based Center for Digital Government: "The cost and time savings that online transactions offer make it a win-win situation for government agencies and citizens."

to government offices), increase social inclusion, and save money, for both users and government (see Case 2).

A number of triple-bottom-line signature initiatives by mainstream ICT companies were listed in the *BusinessWeek* supplement mentioned earlier, and were described at the Digital Dividend conference (*BusinessWeek* 2000). As noted, they included significant initiatives by Hewlett-Packard, Real Networks, and others. We have selected just one example of a mainstream ICT business to profile: 3Com, a company that has exhibited innovative triple-bottom-line thinking both in its mainstream business and in its signature initiatives (Case 3).

From the Digital Dividend commentary and Cases 1 to 3, we can observe the following:

- There are new possibilities for creating social, ecological, and economic gain through application of ICTs that simply would not have been possible before the advent of the technologies. We might describe examples such as GreaterGood.com and e-government initiatives as "pure play" triple-bottom-line propositions.
- There are ways in which mainstream ICT firms can create social, ecological, and economic gain through *both* business practices and Digital Dividend signature initiatives. We might describe examples such as 3Com as

Case 3

Stakeholder engagement and triple-bottom-line thinking at 3Com

In the early 1990s, 3Com found that chlorinated solvents for cleaning its printed circuit boards were becoming increasingly scarce. With President George H.W. Bush accelerating the Clean Air Act Amendments of 1990, 3Com, like many of its competitors, began looking for alternative cleaning processes. It pioneered the implementation of a chlorofluorocarbon-free semi-aqueous cleaning process and, as a result, CEO Eric Benhamou won the President's Environment and Conservation Challenge in 1992. Thus began 3Com's leadership in matters of corporate social responsibility, with many innovative initiatives that became the cornerstone of its growth, leading them to report revenues of $1.5 billion in 1999.

There is no doubt that 3Com is a company with an interesting track record of being innovative in addressing social, economic, and environmental issues. In 2000, the company announced a Digital Divide initiative: a $1 million grant to ten major US cities to help low-income families gain access to the Internet. By 2002, the Urban Challenge Grants totalled $4 million, with $100,000 available in 3Com systems and services across forty US cities. Today all community investments made by 3Com must contribute to bridging the digital divide.

Just as innovatively, in late 2000, 3Com launched Planet Project, a Web-based global poll to gather information on "what it is like to be human at the beginning of the century." One of the most interesting aspects of the poll was that 3Com took full account of the digital divide and addressed social and cultural diversity issues by dispersing thousands of people across the globe to gather results even in areas without Internet access. Approximately 1.2 million people from over 250 countries participated. The political section of the poll received nearly 25,000 responses, 19,000 of them from the United States. The largest numbers of responses from outside the US came from Canada, Australia, Mexico, the United Kingdom, Norway, Germany, Brazil, Chile, Spain, Argentina, Colombia, Denmark, France, and Italy (http://www.3com.com).

"walking the talk" business propositions – uniting technical innovation and good compliance practices to improve social and environmental performance of the business itself (internally) while also embracing triple-bottom-line signature initiatives (externally).

We will now explore the shadow side of ICTs for social, ecological, and economic value propositions.

Reflecting on the Negative

Notwithstanding the positive possibilities described above, there are no guarantees that companies employing digital technologies will *a priori* bridge

Case 4

Home deliveries or a trip to the local grocer?

Webvan is a high-profile business-to-consumer enterprise that has been examined with respect to its environmental, social, and economic performance (Galea and Walton 2002). Webvan is an online California grocery operation supplying home deliveries to nine cities and over half a million customers, with an average transaction value of over US$100. The company received over US$1 billion in venture capital funding.

The conclusions of the Webvan study were not encouraging from a sustainability perspective. Despite obtaining record-breaking levels of venture capital funding, by March 2001 Webvan's losses were running at an annualized rate of US$284 million and the company's share price was just 12.5 cents – down from $25 at its initial public offering in November 1999. The company needed to increase daily orders by almost 50 percent to break even. By 23 July 2001, according to an article by Claire Saliba in *ecommercetimes,* Webvan had attracted two class action suits for misleading investors during the IPO process.

Contrary to the cited assertions of organizations like the World Resources Institute, Galea and Walton (2002) are skeptical that the home delivery of groceries will generate significant social or environmental value. They point to negative environmental impacts such as excess packaging, land use, and significantly lower transportation efficiency (based on a notional 10,000-home customer base and round-trip radii of five and sixty miles for conventional and Webvan deliveries, respectively). Social impacts might be more mixed, with enhanced free time for the wealthy being offset by negative community impacts and other social exclusion factors. The authors of the study noted that Webvan would be out of business before the end of 2001, a prediction that, sadly, proved to be true.

the digital divide and create sustainable value for all stakeholders. Hardware companies had a difficult time in the first few years of the twenty-first century. Software firms and those totally dependent on digitally based products, such as those in entertainment and "business-to-consumer" (b-to-c) e-commerce have not always made profits in the conventional sense. And on the evidence of the Nasdaq high technology stock price index and the significant job cuts announced by many of the industry's best names in recent years, not all digitally based companies added either economic value for their investors or social value for their current and former employees in that period (Wheeler et al. 2003). So the question remains: just what is (or what might be) a sustainable ICT enterprise?

Case 4 shows that even the best funded companies in the e-commerce world may struggle to achieve economic sustainability. Moreover, they may also fail the test of environmental and social gain hitherto assumed by some commentators.

In somewhat pessimistic vein, the British think tank Forum for the Future released a report in January 2001 entitled *Digital Futures: Dot-com Ethics, Business and Sustainability,* which concluded that in general "e-business has failed to grapple with sustainability" (Wilsdon 2001). The Forum for the Future report related the apocryphal tale of the negative environmental impacts of the launch of the third Harry Potter novel by Amazon.com in the US, and bemoaned the apparent lack of understanding of sustainability concepts by many in the ICT sector, who appear to be unaware of the environmental and social impacts of their industry.

Case 5

Successful e-commerce with a social cost

Given the undoubted potential of the Internet to provide social benefit, it is a major disappointment that the most successful commercial activity on the medium to date is pornography. Legal online pornography is the first consistently successful e-commerce product, with earnings estimated at over US$1 billion a year by 2000. Of significantly greater concern to the majority of citizens than soft porn is that part of the industry that is beyond the law in most jurisdictions and that carries a distressing social cost: the manufacture and Internet trade in child pornography.

In March 2001, the Canadian government announced plans to introduce some of the world's toughest legislation to combat child pornography on the Internet. The new law, which received Royal Assent in June 2002, makes it a crime not only to produce or transmit indecent images of children but also to access them (http://canada.gc.ca). It was already illegal in Canada, as well as in the United States and many other countries, to possess child pornography by downloading it from the Internet, but the Canadian legislation is believed to go further than any other by outlawing the very act of viewing child pornography. Prosecutions for transmitting and accessing child pornography could result in prison terms of up to ten years. The law also allows prosecutions in Canada for offences committed abroad.

A complementary approach to the problem is that advocated by the British ICT industry-funded body Internet Watch Foundation (IWF), which was launched in 1996 by PIPEX founder Peter Dawe to address the problem of illegal material on the Internet, particularly child pornography. IWF received initial funding from the European Union, and organizations represented on the IWF Funding Council have included America Online, BT, Cable & Wireless, Energis, Microsoft, NTL, and Yahoo! (http://www.iwf.org.uk). IWF claims a halving of the percentage of Internet pornography traced to Europe during 2003 compared with previous years.

And notwithstanding over a thousand cases and good news stories from the WRI, even the ideal of bridging of the digital divide in the developing world is not without its paradoxes and challenges. Practising what it preaches, the WRI initiated an e-dialogue on "Technology Globalization and the Poor" in November and December 2004 (Paul 2004). The dialogue attracted a range of participants and explored a number of the tensions around achieving the "digital dividend" in the developing world. A number of commentators explained that addressing broader questions of cultural context, equity, and economic development were vital to harnessing the empowering nature of ICTs in the global south. These concerns were also evident in the somewhat skeptical civil society daily Web postings reflecting the debates at the World Summit on the Information Society held in Geneva in December 2003 (Daily Summit 2003).

Ironically, perhaps the most controversial social phenomenon associated with the Internet is probably its most reliable "b-to-c" commercial product: pornography. Whether this phenomenon is a social good or ill is not the point at issue here. There is, however, a shadow side to the phenomenon (and the technology itself) that even the most liberal commentator would not seek to defend: illegal pornography, specifically the trade in child pornography. Our fifth case explores this troubling issue and the legal and voluntary options that are available to governments, judicial authorities, and industry with respect to it (see Case 5).

From Cases 4 and 5, we can observe the following:

- "Pure play" value propositions (i.e., those wholly enabled by ICTs, such as business-to-consumer e-commerce) may fail to realize promised social, ecological, or economic gains. Indeed, they may actually cause significant damage in one or more dimensions, with child pornography being one of the starkest and most odious examples.
- Bridging the digital divide in the global south may face the same cultural, social, and equity challenges as other technologies have faced in achieving "pro-poor" development.
- Based on the commentary of some critics, "walking the talk" business propositions (i.e., those embracing both internal and external triple-bottom-line practices) may not be the norm in the ICT industry.

We now turn to consideration of the synergy between two theoretical constructs (sustainability and social capital) that we believe will help inform future policy options for the ICT sector in Canada.

Sustainability, Social Capital, and the ICT Sector

There is a consensus among large firms that are members of the World Business Council for Sustainable Development that corporate responsibility should be defined in three dimensions: the financial, the social, and the environmental (WBCSD 2002). This is consistent with Elkington's thinking (1998) on the triple bottom line, which calls for firms to achieve balanced progress on economic development, environmental quality, and social justice (or equity). Of these, it is probably the social dimension that is the least developed in terms of corporate strategy and practice (Watts and Holme 1999). It is principally in the social and economic dimensions that problems of the digital divide reside. But if we unwrap the social dimension a little, we discover that it is possible to construct models and frameworks that may help categorize and explore companies' attitudes toward society and the broader concept of sustainability.

Wheeler and colleagues (2003) offer a model for categorizing corporate culture with respect to orientation to stakeholders in service of economic, social, and ecological value (Figure 5.1). The three cultures described are: (1) compliance, (2) relationship management, and (3) the sustainable organization. This model may be helpful in our exploration of the role of ICTs in society in that it spans the breadth of philosophical debate around the purpose of business and reflects the two most often posited assertions for corporate responsibility: "do no harm" and "do maximum good." The model simply restates these two assertions in the three dimensions of sustainability and places them in Level 1 and Level 3, respectively, with an intermediate level (Level 2) as an interface.

The model implies that highest-order sustainability cultures transcend compliance and or "relationship management," phenomena typically associated with conventional approaches to business and corporate citizenship. Sustainable organization culture (beliefs, mindsets, and behaviours consistent with principles of sustainability) and active creation of value for all stakeholders (economic, social, and ecological) implies an approach to relationship building that is generative and open-spirited. It therefore overlaps significantly with the construct of social capital – a term that also deals with stakeholder value – linked to structures, relationships, and beliefs.

The concept of social capital has gained significant attention from theorists and practitioners across the social sciences. Social capital as defined in this chapter refers to the contributions that aspects of social structures and relationships can make to the creation of valued outcomes. Putnam (1993, 1995, 2000), one of the modern founders of the term, refers to social capital as "networks, norms and social trust that facilitate coordination and cooperation for mutual gain." This chapter expands on that definition by including individual gains in addition to mutual gain, but highlighting that *sustainable* individual gains are created through win/win strategies that meet the needs of all stakeholders in the long term.

Nahapiet and Ghoshal (1998) proposed three dimensions of social capital – structural, relational, and cognitive. Networks or interaction patterns that

Figure 5.1

Framework for classifying organizational cultures with respect to the creation and distribution of economic, social (stakeholder), and ecological value

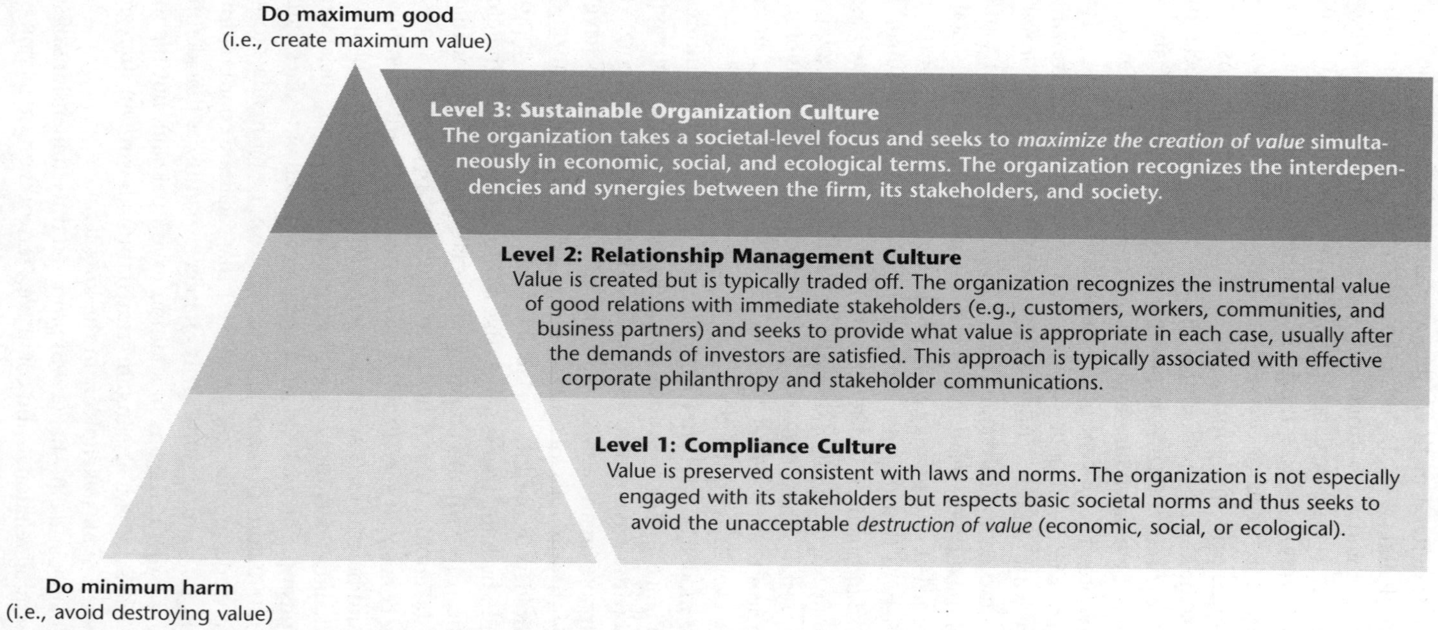

Do maximum good
(i.e., create maximum value)

Level 3: Sustainable Organization Culture
The organization takes a societal-level focus and seeks to *maximize the creation of value* simultaneously in economic, social, and ecological terms. The organization recognizes the interdependencies and synergies between the firm, its stakeholders, and society.

Level 2: Relationship Management Culture
Value is created but is typically traded off. The organization recognizes the instrumental value of good relations with immediate stakeholders (e.g., customers, workers, communities, and business partners) and seeks to provide what value is appropriate in each case, usually after the demands of investors are satisfied. This approach is typically associated with effective corporate philanthropy and stakeholder communications.

Level 1: Compliance Culture
Value is preserved consistent with laws and norms. The organization is not especially engaged with its stakeholders but respects basic societal norms and thus seeks to avoid the unacceptable *destruction of value* (economic, social, or ecological).

Do minimum harm
(i.e., avoid destroying value)

Source: Wheeler et al. 2003, after Stein 1985.

develop among people over time allow connections to be made and bridges to be built among previously unconnected people. These bridges or connections can enable access to and mobilization of resources for individual or mutual benefit. In addition to structural properties of interaction, relational and affective qualities such as trust and reciprocity can facilitate the creation of value by improving willingness to undertake transactions and mutual action. Social capital also refers to the shared mindsets or acquisition of mutual understanding that exists among people and helps to produce value by enabling people to act and reducing time spent on clarifying interests or preferences.

Economic approaches to production focus on use of physical resources and labour in value creation that emphasizes short-term gains for individuals. We argue that sustainable development is an inclusive *societal-level* construct that draws on multiple forms of resources, including social and human capital, and emphasizes broad-based value creation in the long term. Economic theories tend to focus primarily on efficiency and productivity. Social capital theory draws attention to the role that social structures and relations play as facilitators of both. Physical resources are transformed into outputs through labour (that is, people), and therefore social capital plays a role in all economic activities. Social structures and the quality of relations or culture can either facilitate or undermine the production process. Economic activity is inherently social, so creating a social context that facilitates production is a core activity for all successful managers.

Burt (1997, 2000), an influential social capital theorist in the organizational context, has defined social capital in terms of creating bridges across the structural holes in networks. In other words, individuals can connect one person who needs a resource to another who can supply it. In this way, production is facilitated in obvious ways by eliminating impediments to production or by facilitating entrepreneurial activity. Many other social capital theorists (see Adler and Kwon 2002 for a valuable review) have demonstrated how qualitative aspects of social relations, such as levels of trust or friendship, help to motivate people and mobilize resources. A social structure with well-developed networks may have little impact if participants are not willing to engage or share their information, knowledge, and resources in the production process.

More broadly speaking, at the level of communities and nations, social capital is a competitive advantage or disadvantage. As Putnam (2000), Fukuyama (1995a, 1995b), and Woolcock (2000) have argued, economic development is significantly affected by the extent of interaction among citizens and the levels of societal trust that exist. In a public policy context, therefore, effective social and international development spending should be viewed as investments in sustainable economic development rather than costs that detract from economic profits.

Social capital is therefore potentially a critical component in both explaining *and achieving* sustainable development. Social capital theories require us to include social resources and impacts in our conceptions of production, drawing attention to the need to create broad-based value for both the focal organization and its stakeholders. Inevitably, social capital impels us to view economic growth as integrally connected to the social context of the people who participate in and are affected by it. And because of the highly networked and relationship-dependent nature of the sector, there can be few industries of greater interest to social capital and sustainable development theorists than those reliant on ICTs.

We now turn to our preliminary research on Canadian ICT organizations and their attitudes toward both sustainability and social capital.

The Canadian ICT Sector

Despite a temporary downturn during the first two years of the twenty-first century,[4] in 2003 the Canadian information and communication technology sector still represented over 5 percent of the Canadian economy, with a $54.5 billion contribution to GDP (Government of Canada 2005). Historically, the sector has been one of the fastest-growing elements in the economy, and between 1997 and 2003 represented 11.1 percent of all GDP growth in Canada (average 9 percent growth per annum for the sector itself). In 2001, research and development (R&D) expenditures in the ICT sector were estimated at $5.3 billion, representing no less than 45.8 percent of total Canadian private sector R&D. From 1990 to 1997, R&D expenditures in the ICT sector had a compound annual growth rate of 9.1 percent, compared with 7.5 percent in the total Canadian private sector, representing almost 42 percent of total Canadian private sector R&D (ITAC 2002).

The ICT sector in Canada consists of a diverse range of business organizations in telecommunications, cable, and information technology manufacturing and services (ITAC 2002; Government of Canada 2002a). Globally, the ICT sector is dominated by large international firms with many international subsidiaries, and a few high-profile firms focusing on domestic markets. This structure poses special challenges for a smaller and more dependent economy such as Canada, which has traditionally been dominated by the US in technology-based industries. The pace of change in the last ten years has also meant that large incumbent US companies have tended to dominate developments more effectively through their ability to drive mergers, acquisitions, and divestments.

Notwithstanding the dominance of US firms, the Canadian government has recognized the importance of this sector to the overall competitiveness of the domestic economy, and since the early 1990s has been investing in the development and dissemination of information technologies, both directly through research grants and tax credits and indirectly through provision of necessary infrastructure to facilitate the emergence of and innovation among ICT companies (Government of Canada 2002a).

Judging by the performance statistics on growth and research and development noted above, government efforts aimed at facilitating the growth of the Canadian ICT sector have certainly not harmed economic performance. The sector has also been successful in terms of generating foreign direct investment: since the mid-1990s, ICT products have accounted for, on average, about 30 percent a year of total investment spending in Canada, raising the share of ICT in the capital stock to 4.5 percent.

These statistics, together with commentaries by the Government of Canada (2002a, 2005), might be interpreted to mean that Canada's ICT sector is thriving. In reality, however, although the sector appears to be positioned strongly compared with Canada's other knowledge-based and traditional manufacturing industries, its performance vis-à-vis US and European technology industries does not appear so robust.

Concerns about possible underperformance in the sector are not new (Himmelsbach 1999). A comparative overview of the Canadian ICT sector with respect to its international counterparts in the developed world – for example, on the innovative capacity of ICT firms, their investments in R&D, and their overall trade position – reveals that Canada has lagged behind most G7 nations in almost all key measures (OECD 2000). According to the OECD *Information Technology Outlook,* Canada has historically been outperformed by other industrialized nations. In the period 1980-97, Canada's value added to GDP averaged 2.9 compared with the US ICT sector's contribution of 4.4 percent. As a result, from an early stage Canada maintained a negative trade balance in ICT manufacturing and services (18 percent deficit in 1997), despite the fact that it exports over three-quarters of its overall production. Canada's trade deficit in ICT products and services increased 67.7 percent from 1993 to 2000, when it stood at over $20 billion.

The underperformance of the Canadian ICT sector is largely attributable to its small size compared with other industrialized countries, its relative slowness in adopting new technologies, as well as its lower investment in innovation. These trends are also reflected in the positioning of individual firms within the Canadian ICT sector. There are no Canadian firms among world's top fifty ICT companies.

Sectoral Interviews

As part of the multi-stakeholder Sustainable Canada initiative led by York University, and in order to explore Canadian ICT sector attitudes toward sustainability and questions of social capital, eight in-depth interviews were conducted during the first six months of 2002. Interviewees for this exploratory study were drawn from a range of Canadian-based ICT firms and organizations as well as independent consultants with substantial experience in the industry. The firms contacted to participate in the study were a mix of Canadian and non-Canadian companies: Bell Canada, Gennum, IBM, ITAC, Lucent Canada, Microsoft Canada, Nortel Networks, and Telus. The Information Technology Association of Canada (ITAC) and two industry consultants were also contacted. The majority of interviewees were senior managers with extensive corporate experience in sustainability issues.

Results of Interviews:
Possibilities for Sustainability-Focused Collaboration

In our interviews, sustainability was seen as a somewhat marginal rather than core issue for the sector. Responses to the question of whether sustainability was a frame of reference in day-to-day operations were quite varied, and on average suggested that it was not (with only small disagreement). On the other hand, most interviewees felt that it was important for their organization to be recognized as a leader in sustainability.[5]

The ICT sector sees sustainability in two ways: internal to the industry and external. Internally, issues focus on how the sector needs to deal with its own sustainability through reduction of waste and improved human development. On the other hand, the ICT sector views itself as an "enabler" of greater sustainability in other industries through the supply of technology that improves the efficiency of resource use, reduces the demand for travel, and is helping to "dematerialize" the economy. On the social side, technology makes knowledge more widely available than it was. Overall, most insiders we interviewed saw ICT as a positive force for encouraging sustainability rather than as a contributor to environmental and social problems.

The Canadian ICT industry is divided between manufacturers and service/software providers, and sustainability was viewed quite differently from these perspectives. Manufacturers were more concerned about environmental impact such as waste disposal and efficient use of resources, while the non-manufacturing side tended to view sustainability in terms of social issues. Organizations involved with manufacturing recognized that there were significant environmental concerns related to hazardous waste and disposal of products. Thus the industry association ITAC identified waste and recycling as one of its three priorities and convened a working group to examine the issues with a view to developing a national voluntary program that would see the industry take responsibility for disposal of its products and preferably engage in recycling. Clearly the issue is extremely complex, with a broad range of equipment involved and differing regulatory regimes in different provinces and municipalities as well as potential issues regarding cross-border disposal. This effort was partially driven by proposed "product take back" legislation that was tabled in Ontario and Manitoba and that was being considered in British Columbia.

The digital divide was also identified as a sustainability issue but was not as broadly recognized by our interviewees. The digital divide was seen as a philanthropic issue to a large extent, although, as the National Broadband Task Force notes (Government of Canada 2002b), this issue could translate into increased demand in the industry.

Particularly in light of the downturn in the industry in 2001-2002, it was widely agreed that organizations must focus on issues that will contribute to the economic bottom line in some way, for example, reduce costs or increase demand. Most interviewees, however, felt that the profile of sustainability was increasing in the industry, and although legislation was not preferred, sustainability-inspired proposals were putting pressure on the industry to examine the issue. Another potential driver of change identified was internal, from employees, who for the most part are well educated, highly skilled, and concerned about the ethical stance of the organizations they work for. A final incentive to move forward was the fact that Europe is generally perceived as being significantly ahead of Canada in making progress on these issues. Canada was seen as a relative leader in North America, while the US was seen as lagging in recognizing and dealing with sustainability issues.

While sustainability may not be at the top of ICT industry priority lists, the industry has some potential advantages that could help it move forward on such issues. One is that its core identity includes an association with innovative problem solving. This can-do attitude could be a significant asset. In addition, despite the disruptions in the structures of the industry in 2001 and 2002, most interviewees still indicated that the industry works well together on matters of mutual interest.

The industry has substantial social capital. Members of the industry indicated that they interact frequently at associations, conferences, and industry task forces. Many others reported strong informal ties, particularly among those in similar functions. While responses regarding levels of trust were quite varied, most agreed that there was positive trust among members of the industry, particularly long-standing members. While there was quite a bit of variation in attitudes, many felt that the industry was a "tightly knit" community and reported well-developed networks among organizations for coordinating activities. As a naturally "networked" industry, agreeing on standards and coordinating activities is a core competence.

Some of the larger, more visible companies saw licence to operate (tier one in Figure 5.1) as an increasingly important issue, and being seen as a leader on sustainability issues may be beneficial on this front. Despite these advantages, there was concern that competitive pressures and the immediate need for survival may overshadow issues that cannot be linked directly to the bottom line in the short term. As noted, Canada was not seen as an international leader in the ICT sector, where some European companies such as Ericcson were seen as far ahead of North American firms. On the other hand, Canada is recognized by many internationally for spearheading global efforts to reduce chlorofluorocarbons (CFCs), although awareness of this accomplishment may not be as widespread in Canada. Another concern was the absence of pressure from customers. This may change in the future, as economic pressures in the industry push customers' interests (including European customers' interests) to the forefront and "Design for Environment" (DfE) is increasingly seen as important by those clients.

The approach that most interviewees felt was necessary for progress on any strategic Canadian initiative on sustainability for the ICT sector was one that was quite formal, included strong leadership from dominant players in the industry, and partnered with governments, especially the federal government. The ICT sector has a strong relationship with government and feels that the federal government's innovation agenda is quite compatible with their industry objectives. Members of the industry thought it very important not to overstate Canada's position with respect to sustainability. Many respondents also expressed a reluctance to "stick your neck out," as some stakeholders might be willing to "chop it off." These comments suggest a desire for a balanced and authentic approach that will withstand the scrutiny of potentially critical stakeholders.

As noted earlier, members of the ICT industry strongly identified with the industry and saw themselves as innovative problem solvers. They also identified with technology, change, and dedication to their customers. Most felt that the identity of the industry was quite homogeneous. While few identified sustainability as an aspect of their core identity, most agreed that being seen as leaders on sustainability was important for their organizations.

The sector's identity has shifted in the past ten years toward a more dynamic, customer-driven one, reflecting the industry's shift from monopoly to competition. Relations with stakeholders were thus seen as very important; a broad range of external stakeholders were mentioned. Organizations viewed the communities they served, governments, shareholders, and customers as key stakeholders. Few non-governmental or pressure groups were mentioned: the industry experiences little concerted pressure toward sustainability from pressure groups or other external stakeholders. Employees were seen as important internal stakeholders, particularly given the importance of hiring and retaining highly skilled staff across the industry. In fact, one respondent identified access to skilled employees as the core sustainability issue for the industry. Employees were also seen as valuing sustainability and working for an organization that was "doing the right things."

In summary, the innovative, problem-solving identity of members of the sector and the high value placed on sustainability by employees were – and remain – key sources of leverage for strategic sustainability initiatives in the sector. The absence of significant pressure from external stakeholders was viewed as one reason the sector has not moved forward more quickly on sustainability issues to date. Thus, our preliminary research enables us to make the following observations:

- Opinion formers in the Canadian ICT sector believe in the inherent potential for the ecological and social sustainability of the sector. They also recognize the internal versus external dimensions of the sustainable development agenda for the sector (i.e., employee perspectives and other business issues, plus the need to help bridge the digital divide).
- Opinion formers in the Canadian ICT sector recognize the national competitiveness issues involved in the sustainability agenda, but also feel that there is greater leadership in Europe and that in Canada there is a need for stronger leadership from both government and major players in the sector.
- Opinion formers in the Canadian ICT sector feel that there is potential for concerted action on sustainability in the sector based on the high level of trust already evident, and that this might best be galvanized by an approach linking sustainability to innovation, which in turn links Canadian government policy with industry identity.

Conclusion

In summary, our exploration of cases, our theoretical analysis, and our preliminary research have led us to advance eight observations:

- There are new possibilities for creating social, ecological, and economic gain through application of ICTs that simply would not have been possible before the advent of the technologies.
- There are ways in which mainstream ICT firms can create social, ecological, and economic gain through *both* business practices and Digital Dividend signature initiatives.
- "Pure play" value propositions (i.e., those wholly enabled by ICTs, such as business-to-consumer e-commerce) may fail to realize promised social, ecological, or economic gains.
- Bridging the digital divide in the global south may face the same cultural, social, and equity challenges as other technologies have faced in achieving "pro-poor" development.
- Based on the commentary of some critics, "walking the talk" business propositions (i.e., those embracing both internal and external triple-bottom-line practices) may not be the norm in the ICT industry.

And, as we noted just above from our interviews:

- Opinion formers in the Canadian ICT sector believe in the inherent potential for the ecological and social sustainability of the sector.
- Opinion formers in the Canadian ICT sector recognize the national competitiveness issues involved in the sustainability agenda.
- Opinion formers in the Canadian ICT sector feel that there is potential for concerted action on sustainability in the sector based on the high level of trust already evident, and that this might best be galvanized by an approach linking sustainability to innovation, which in turn links Canadian government policy with industry identity.

Earlier, we described the cognitive, relational, and structural dimensions of social capital. We noted the convergence of social capital theory and sustainability as a construct requiring cultural (cognitive values and beliefs) and stakeholder-inclusive (structurally and relationally) corporate cultures.

Applying our theoretical frameworks to our eight observations, it appears that cognitively (and culturally), there are no intrinsic barriers to ICT firms pursuing sustainability as a sector and/or as individual firms committed to creating economic, social, and ecological value for stakeholders. Of course, there are no guarantees that the sector in Canada will deliver such value, but at least there seem to be no philosophical barriers. In terms of the relational dimension of social capital, despite competitive pressures and structural rearrangements, there seems to be a high level of trust within the sector in Canada. This may further assist in the development of strategically valuable initiatives in sustainability and innovation, led by the sector.

In terms of structures and networks, however, it appears that many of the companies comprising the Canadian ICT sector are strategically oriented to ensure a stable domestic position, rather than to pursue international expansion. Moreover, due to competitive dynamics evident among firms in the sector (principally the domination of the sector by large US companies), it may not be typical for Canadian ICT firms to form strong international networks or long-term alliances of their own, and therefore social capital formation and significant efforts on sustainability are more limited than they might be.

So we might conclude that optimizing social capital within *and beyond* the Canadian ICT sector may require initiatives focused particularly on improving *structural* (i.e., network-related) social capital. Furthermore, such initiatives may be a prerequisite for both greater international competitiveness *and* enhanced sustainability.

As Adler and Kwon's review (2002) of the social capital literature suggests, despite the existence of over twenty different definitions of social capital, most include both structure (i.e., networks) and content (e.g., trust, shared norms, or understandings) of social relations. New technologies have the potential to foster expansion of networks and enable people to develop trust and mutual understanding that will facilitate sustainable development. These new technologies help to overcome barriers of time or geography that previously confined the development of social capital to a local context, allowing certain people or geographic locations to be excluded. There is no industry that understands this better than the ICT sector itself.

As many social capital theorists argue, social capital can be fostered through increased interactions that enable people to know, trust, and understand one another. While the link between interaction and creation of social capital has not been shown to be causal, theoretically high levels of *voluntary* interaction should be indicative of higher levels of social capital. While actors in a particular context may be forced into regular interaction, voluntary interaction should signal the existence of a positive quality to the interaction.

Some social capital theorists (Burt 1997; Rowley et al. 2000) have viewed the existence of dense connections as constraining, and therefore a liability rather than an advantage. This chapter argues that sustainable development *requires* the focal actor to proactively engage and resolve potential conflicts of interest with stakeholders in order to create sustainable economic profits. The increased opportunity for interaction and dialogue that the new technologies offer should enable organizations to engage those both favourably and unfavourably disposed to their enterprise in constructive discussions that can produce greater shared understanding, trust, and support. Alignment among the interests of stakeholders thus enables the focal actor, such as a Canadian ICT firm, to create value across its "value-based networks" (Wheeler et al. 2003).

Not only can individuals interact more often but new technologies offer broader access to different forms of interaction that can enable dialogue rather than a simple information exchange. The opportunity to discuss issues more fully, potentially including visual displays such as those through webcams or videoconferences, which allows for more comprehensive communication, should result in improved understanding between stakeholders and the focal actor. Frequent and more personalized interactions should aid in the development of trust, shared norms, and mindsets that will remove obstacles and encourage constructive participation. Again, the ICT sector knows this better than any other sector, and indeed depends on it for its economic success.

Thus we argue that in terms of (1) the establishment of social licence to operate through effective compliance with laws and societal norms (Level 1 in Figure 5.1), (2) effective management of stakeholder relationships (Level 2), and (3) the generation of triple-bottom-line value for all stakeholders (Level 3), Canadian ICT companies should be in an excellent position to leverage their technological expertise to generate stakeholder support both in home markets and more internationally. Arguably, this is precisely what the signature initiatives of large US-based ICT firms are designed to do: align external stakeholders through inclusion in value-based networks or communities of interest that maximize loyalty and market advantage. The question that remains is whether Canadian-based ICTs can generate similar advantages. We believe that they can – partly because of existing levels of cognitive and relational social capital within the sector, partly because of the natural tendency of Canadian institutions to operate in a stakeholder-inclusive manner, and partly because the nature of the industry and its products are intrinsically conducive to the creation of powerful networks.

With the inclusion of more and more diverse stakeholders in market-based economic processes, the external structural social capital of the Canadian ICT sector has the capacity to grow exponentially. Social capital, like other forms of capital, is often created as a by-product of successful transactions. Capital tends to beget more capital; hence, those with social capital may end up with more while those without may again fall further behind. This phenomenon is central to the competitiveness dilemma for Canadian ICT firms.

Happily, it seems that social capital is often an unintended by-product of interactions conducted for another purpose. Communities that interact a lot tend to maintain the social capital they have and to create more as a by-product of their frequent mutual engagement in activities. Social capital theorists have called this the virtuous cycle (Putnam 2000). Social capital theory has tended to regard social capital as an endowment and has not explored the ultimate sources of social capital. Proximate sources such as homogeneity or high levels of interaction have been suggested as sources, but these are clearly not created but rather are givens. So if Canadian ICT firms are by their nature dependent on (primarily) US networks, US sources of innovation, and US social capital, they could choose to find new ways of building social capital *of their own* in service of greater competitive advantage. Joint initiatives in sustainability and innovation where there are fewer competitive pressures to act as barriers may be one example of how to stimulate a virtuous cycle of social capital creation both domestically and internationally.

We argue that social capital can be created by gradual inclusion of a broader array of participants in interactions. In other words, the social capital created as a by-product of a "sustainability interaction" (for example, with domestic and international partners) may be "seeded" in virgin soil, potentially creating a foothold that enables a once disenfranchised and unconnected small Canadian ICT firm to become a member of a more powerful network and thereby gain the right to appropriate and leverage the broader network's social capital. Further, this firm may be able to use this network affiliation to begin engaging others in that community, expanding the Canadian ICT social capital created via a ripple effect. The virtuous cycle, therefore, should gradually be engaged once some threshold is reached.

One final – potentially very important – factor in the creation of social capital and capacity for sustainability is the *transparency* that new technologies allow with key stakeholders, including investors. This is especially important in the current context of debates about corporate governance and the lack of confidence in capital markets – still worse, the absence of evidence that even indices of good corporate governance practices are reflected in stock price performance (Wheeler and Davies 2004, 2006). Controlling flows of information has become much more difficult as technologies enable fast and cheap dissemination of information. Transparency is of great importance to building trust and understanding, and therefore social capital. With the right strategies in corporate disclosure in place, as more people become aware of processes of decision making, distrust should decline. In the future, organizations that may have been tempted to make a less than ethical decision may hesitate, given the real potential for widespread dissemination of this information. Again, Canadian ICT firms can choose to play a leadership role here.

In order to increase levels of social capital and thus the international competitiveness and sustainability of the Canadian ICT sector, some practical proposals arising from this analysis may be proposed. They focus especially on creating internal and external synergies and the need to build the structural component of social capital in service of economic, social, and ecological gain. For example:

The Canadian ICT industry could lead in more actively disseminating best practices on the internal dimension of "walking the talk" sustainability initiatives in ICT companies. We have already noted the earlier reluctance of some Canadian-based ICTs to be too bold in advancing an image when there may be disconnects between external signature-type gestures and internal practices. ITAC, however, has created an infrastructure for addressing internal sustainability management issues that is national in scope and thus allows for the development of a broad-based governance structure for broader issues too. This effort might include proactiveness in reporting and environmental management, and may be positioned as building on the previous successes of the CFC elimination effort. Should the solution include any technical advances, these may be used to broker international relationships with others seeking to solve similar problems, as the CFC campaign did. Partnership with Industry Canada through its Innovation Strategy may create the necessary link to government, identified in our interviews as an important structural dimension.

Partnering between hardware and software/services firms, civil society organizations, and governments on domestic and international (i.e., external) "walking the talk" initiatives led by individual Canadian-based ICT businesses could be encouraged. Sociological literature suggests that a proactive move by a dominant industry player can create a bandwagon effect among less dominant players. Thus there is a real opportunity for a dominant Canadian player to take a leadership role, for example, in bridging the digital divide, and reap reputational advantage as well as potentially create demand for content and other applications among the expanded user base over the long term. There is no reason why, for example, hardware providers such as Nortel Networks or Research in Motion (RIM) should not identify leading Canadian-based software suppliers – including those with US parents – with whom to partner in ambitious "walk the talk" initiatives. If such initiatives are made more strategic through partnerships with governments, economic development agencies, and civil society organizations, so much the better in terms of external advantage.

The Canadian government (e.g., Industry Canada and the Canadian International Development Agency) could lead the brokering of partnership arrangements between the Canadian ICT sector and specific regions in both developed and developing countries, involving civil society organizations and local governments. It is unlikely that the Canadian ICT sector can establish strategic partnerships aimed at bridging the digital divide globally (as has been HP's intention). A more selective approach, however, based on regions where Canada and Canadians have historic ties (i.e., pre-existing social capital), may be appropriate in forging collaborative and mutually beneficial partnerships. Twinning the sustainable development and innovation agendas may prove particularly appropriate in harnessing the power of commercial foreign direct investment, overseas development assistance, and direct commercial (e.g., supply chain) arrangements designed to deliver social, ecological, and economic value.

In each case, the net impact of these recommendations would be to build social, ecological, and economic value for Canadian ICT firms and their stakeholders. There would, however, be an assumption of radically broadening the horizon and breadth of stakeholders in a more broadly based dimension of structural social capital in order to augment the existing cognitive and relational strength of Canadian ICT firms with respect to internal and external sustainability.

Acknowledgments

This chapter is dedicated to the memory of Michael Austin Perkin (1957-2002), who contributed significantly to the ideas expressed here and whose enthusiasm for the potential societal benefits of ICTs was boundless.

The chapter was researched and written with the assistance of SSHRC Initiative on the New Economy Grant Number 510189 on The Sustainability of Canadian Business in the New Economy: "Sustainable Canada" (with contributions from Alice Koldertsova and Adrian Capobianco).

Notes

1 In 1998, the OECD countries developed an industry-based definition of the Information and Communications Technologies (ICTs) sector using the International Standard Industrial Classification (ISIC). According to this standard, the ICT sector has been defined as a combination of manufacturing and services industries that capture, transmit, and display, by electronic means, data and information. As a result of the OECD definition, the Canadian information and communications technology sector is defined as consisting of ICT manufacturing firms such as computer equipment, communications, wire, and cable producers, and ICT intangible and goods-related services, including software and computer and telecommunications programs.

2 In this chapter, we define the "new economy" in the broadest sense, to reflect the importance of leveraging new competencies, knowledge, and technology (including information and communication technology) in order to sustain competitiveness and business success in the global economy. In this context, we assert the links between concepts of sustainability, corporate social responsibility, and business strategy, and the abilities of digitally based firms to create economic, social, and environmental value for stakeholders in the long term. Part of this value creation may enable the bridging of the so-called digital divide.

3 See http://www.nextbillion.org for a comprehensive set of commentaries and case resources on the "digital dividend" as it applies to international development and poverty reduction.

4 Between its peak in the third quarter of 2000 and the end of 2002, the ICT manufacturing industries lost over 40 percent of their output.

5 In 2002, the Global e-Sustainability Initiative released a report specifically reviewing the role of ICTs in sustainability, which was consistent with this perception.

References

Adams, S. 1997. *The Dilbert future.* New York: HarperCollins.

Adler, P.A., and S.-W. Kwon. 2002. Social capital: Prospects for a new concept. *Academy of Management Review* 27 (1): 17-40.

Burt, R.S. 1997. The contingent value of social capital. *Administrative Science Quarterly* 42 (2): 339-65.

–. 2000. The network structure of social capital. *Research in Organizational Behaviour* 22: 345-423.

BusinessWeek. 2000. Supplement. Is the "digital divide" a problem or an opportunity? *BusinessWeek,* 18 December.

Castells, M. 2000. *The rise of the network society.* Malden, Uk: Blackwell Publishers.

Daily Summit. 2003. Unlocking WSIS for the world. http://ww.dailysummit.net/english/archives/categories/digital_divide/index.asp.

Dunn, S. 2000. Micropower: Electrifying the digital economy. *Greener Management International* 32: 43-56.

Elkington, J. 1998. *Cannibals with forks: The triple bottom line of 21st century business.* Gabriola Island, BC: New Society Publishers.

Environmental Protection Agency. 2002. *Safe drinking water information system.* Washington, DC: USEPA Center for Information and Statistics.

European Information Technology Observatory. 2002. *The impact of ICT on sustainable development.* http://www.eito.com (accessed 23 August 2003).

Fukuyama, F. 1995a. *The great disruption: Human nature and the reconstitution of social order.* New York: Free Press.

–. 1995b. *Trust: The social virtues and the creation of prosperity.* London: Penguin Books.

Galea, C., and S. Walton. 2002. Is e-commerce sustainable? Lessons from Webvan. In *Ecology of the new economy: Sustainable transformation of global information, communication, and electronics industries,* ed. J. Park and N. Roome. Sheffield, UK: Greenleaf.

Global e-Sustainability Initiative. 2002. Sector report to the World Summit on Sustainable Development. In *Information and communication technology sector report.* http://www.gesi.org (accessed 3 January 2003).

Government of Canada. 2002a. Sector profile. In *Sector competitiveness framework report: Telecommunications equipment industry, software and computer industries*. http://www.strategis.ic.gc.ca.
–. 2002b. *National Broadband Taskforce*. http://www.broadband.gc.ca/about/nbtf-about_e.asp.
–. 2005. Industry Canada ICT sector profile. http://www.strategis.ic.gc.ca.
Handy, C. 1998. *Beyond certainty: The changing world of organization*. Boston: Harvard Business School Press.
Hegel, J. III, and A.G. Armstrong. 1997. *Net gain: Expanding markets through virtual communities*. Boston: Harvard Business School Press.
Hill, K.A., and J.E. Hughes. 1998. *Cyberpolitics: Citizen activism and the age of the Internet*. Lanham, MD: Rowman and Littlefield.
Himmelsbach, V. 1999. ITAC president complains Canada still "way behind." *Computing Canada* 25 (22): 25-26.
Information Technology Association of Canada (ITAC). 2002. *Annual report 2001-02*. Ottawa: ITAC.
International Labour Organization (ILO). 2001. *World employment report 2001: Life at work in the information economy*. Geneva: ILO.
McGovern, G. 1999. *The caring economy: Business principles for the new digital age*. Dublin: Blackhall.
Moss Kanter, R. 2001. *Evolve: Succeeding in the digital culture of tomorrow*. Boston: Harvard Business School Press.
Nahapiet, J., and S. Ghoshal. 1998. Social capital, intellectual capital and the organizational advantage. *Academy of Management Review* 23 (2): 242-66.
Organisation for Economic Co-operation and Development (OECD). 2000. *Information technology outlook: ICTs, e-commerce and the information technology*. Paris: OECD.
Park, J., and N. Roome, eds. 2002. *The ecology of the new economy: Sustainable transformation of global information, communication and electronics industries*. Sheffield, UK: Greenleaf.
Paul, J. 2004. *Technology globalization and the poor. Summary of the Global Knowledge for Development Virtual Conference*. Washington, DC: World Resources Institute.
Paul, J., R. Katz, and S. Gallagher. 2004. *Lessons from the field: An overview of the current uses of information and communication technologies for development*. Washington, DC: World Resources Institute.
Post, J.E. 2000. Moving from geographic to virtual communities: Global corporate citizenship in a dot.com world. *Business and Society Review* 105 (1): 27-46.
Prahalad, C.K., and A. Hammond. 2002. Serving the poor, profitably. *Harvard Business Review* 80 (9): 48-57.
Prahalad, C.K., and S.L. Hart. 2002. The fortune at the bottom of the pyramid. *Strategy + Business Magazine* 26: 2-14.
Putnam, R.D. 1993. The prosperous community: Social capital and public life. *American Prospect* 13: 35-42.
–.1995. Bowling alone: America's declining social capital. *Journal of Democracy* 6 (1): 65-78.
–. 2000. *Bowling alone: The collapse and revival of American community*. New York: Simon & Schuster.
Reich, R.B. 1992. *The work of nations*. New York: Vintage.
–. 2001. *The future of success*. New York: Knopf.
Reichheld, F., and P. Schefter. 2000. E-loyalty: Your secret weapon on the Web. *Harvard Business Review* 78 (4): 105-13.
Rifkin, J. 1995. *The end of work*. New York: Putnam.
Romm, J., A. Rosenfeld, and S. Hermann. 1999. *The Internet economy and global warming*. Arlington, VA: Center for Energy and Climate Solutions.
Rowley, T., D. Behrens, and D. Krackhardt. 2000. Redundant governance structures: An analysis of structural and relational embeddedness in the steel and semiconductor industries. *Strategic Management Journal* 21 (3): 369-86.
Sheats, J.R. 2000. Information technology, sustainable development and developing nations. *Greener Management International* 32: 33-41.

–. 2001. Information technology in sustainable development. In *Technology, Humans and Society,* ed. R. Dorf, 146-58. San Diego: Academic Press.
Silicon Valley Environmental Partnership. 1999. *Environmental index.* http://www.svep.org.
Smith, C.W. 2002. *Digital corporate citizenship: The business response to the digital divide.* Indianapolis: Indiana University Center on Philanthropy.
Stiglitz, J.E. 2002. *Globalization and its discontents.* New York: W.W. Norton.
Tapscott, D., and D. Ticoll. 2003. *The naked corporation: How the age of transparency will revolutionize business.* New York: Free Press.
Tapscott, D., D. Ticoll, and A. Lowy. 1999. *Digital capital: Harnessing the power of business webs.* Boston: Harvard Business School Press.
University of Texas. 2002. *Zapatistas in cyberspace: A guide to analysis and resources.* http://www.eco.utexas.edu/faculty/Cleaver/zapsincyber.html.
Watts, P., and R. Holme. 1999. *Meeting changing expectations. Corporate social responsibility.* Geneva: WBCSD. Online: http://www.wbcsd.org/publications/csrpub.htm.
Wheeler, D., and R. Davies. 2004. Gaining goodwill: Developing stakeholder approaches to corporate governance. *Journal of General Management* 30 (2): 51-74.
–. 2006. Why corporate governance rankings do not predict future value: Evidence from Toronto Stock Exchange listings 2002-2005. In *Corporate governance and sustainability: Challenges for theory and practice,* ed. S. Benn and D. Dunphy. London: Routledge.
Wheeler, D., and J. Elkington. 2001. The end of the corporate environmental report. Or: The advent of cybernetic sustainability reporting. *Business Strategy and the Environment* 10: 1-14.
Wheeler D., B. Colbert, and R.E. Freeman. 2003. Focusing on value: Reconciling corporate social responsibility, sustainability and a stakeholder approach in a network world. *Journal of General Management* 28 (3): 1-28.
Wheeler, D., K. McKague, J. Thomson, R. Davies, J. Medalye, and M. Prada. 2005. Creating sustainable local enterprise networks. *MIT Sloan Management Review* 47 (1): 33-40.
Wilsdon, J. 2001. *Digital futures: Dot-com ethics, business and sustainability.* London: Forum for the Future.
–, ed. 2002. *Digital futures: Living in a dot-com world.* London: Earthscan.
Woolcock, M. 2000. Social capital: Implications for development theory, research, and policy. *World Bank Research Observer* 15 (2): 225.
World Business Council on Sustainable Development (WBCSD). 2002. WBCSD website. http://www.wbcsd.ch/aboutus/ (accessed 3 January 2002).
World Resources Institute (WRI). 2001-2005. Digital Dividend website. http://www.digitaldividend.org.
World Resources Institute (WRI), United Nations Environment Programme (UNEP), and World Business Council for Sustainable Development (WBCSD). 2002. *Tomorrow's markets: Global trends and their implications for business.* Washington, DC: WRI.

6

Innovation, Architecture, and the Changing Role of Design Professionals: Assessing the Ford Model U

Carey Frey

> Design can really be seen as the first signal of human intention.
>
> – William McDonough and Michael Braungart, "The Next Industrial Revolution"

> Design is really just applied foresight.
>
> – Paul Hawken, Amory Lovins, and L. Hunter Lovins, *Natural Capitalism*

A significant barrier to the implementation of sustainable development is the ongoing depletion of the earth's natural resources by entrenched industrial practices that were not originally designed with potential environmental consequences in mind. Modern technologies and revolutionary thinking about a new model for manufacturing now provide a way forward to attaining sustainable production. One of the keys to realizing this goal will be a fundamental reconsideration of traditional approaches to industrial product and process design. This transformation must ultimately occur, however, through strategic leadership in the professions and organizations that are capable of modernizing the industrial system.

The Ford Motor Company has recently demonstrated this type of leadership by taking what may be considered a notable first step toward achieving sustainable production in the automotive sector. At the 2003 Detroit Auto Show, Ford unveiled a new environmentally intelligent concept car called the Model U. Ford's goal is for the Model U to have the same impact on automotive production in the twenty-first century as its Model T did in the twentieth century (Ford Motor Company 2003). Nevertheless, skeptics may easily dismiss the Model U as just a splashy marketing campaign from a company with few sincere intentions about actually developing the product for the mass market. An analysis of the Model U design according to the criteria for a new industrial model will determine whether the concept vehicle is technologically feasible and capable of mass sustainable production.

This chapter will identify and establish criteria for an analysis framework to assess sustainable product design. The framework will be assembled from key concepts evident in the phenomenon of technological innovation in firms, new approaches to holistic system architecture, and the evolving role of the design professions. The chapter will illustrate how firms face significant obstacles to designing both innovative and sustainable products. The critical role of the design process will be analyzed and linked to the increasing politicization of architecture and industrial planning. The potential for sustainable production to drive change in the education and regulation of those professionals responsible for design will also be explored. This framework will then be applied to the case of the Model U in order to evaluate the commitment of the design team to the principles of sustainability and the viability of the end result.

Following the introduction, the analysis framework is developed across a section of three segments comprising innovation, architecture, and the design professions. This section concurrently rationalizes the theory, process, and application of product design by the firm and integrates these concepts with the changing relationship between industry, society, and the environment. The next section presents the case study of the Model U, beginning with a history and overview of the concept vehicle design process, specific green technologies, and other noteworthy technical features. The application of the analysis framework to the Model U follows, with the additional considerations of technological interdependency and the eco-effectiveness of the design. Finally, conclusions reached in the analysis are linked to future prospects of the industrial transition to sustainable production.

An Analysis Framework for Sustainable Product Design

Sustainable Production and Technological Innovation in Firms

The industrial revolution has greatly expanded the material development of humankind, but at a severe price. Since the mid-eighteenth century, more destruction of the environment has occurred than in all prior history, and this rate of loss in the "natural capital" is increasing proportionately to material gains in well-being (Hawken et al. 1999). In this context, sustainability can be understood as an organized response to threats against human and planetary survival (Presley and Meade 2002). Functioning systems must be maintained over time in order to entrust opportunities to future societies regarding the quality of life they desire. "This not only refers to the natural systems underlying all industrial economies but also to social, economic and, in particular, institutional systems" (Spangenberg 2001, 30). The economic system is frequently targeted for radical reform by sustainability advocates due to the behaviour of the consumers, producers, and industrial designers whose actions determine the extent of resource extraction from the environment. Sustainable production proposes the creation of goods and services using systems that are environmentally benign, socially rewarding, and economically viable (Veleva and Ellenbecker 2000).

Since the 1992 Earth Summit in Rio de Janeiro, two influential strategies have emerged and are competing as approaches for achieving sustainable production: eco-efficiency and eco-effectiveness. The report of the World Commission on Environment and Development (WCED), known as the Brundtland Report, first linked eco-efficiency to sustainable development by proposing more efficient resource use within industry in order to minimize the impact of industrial operations on the environment (WCED 1987). Essentially, the term has been coined to mean "doing more with less." William McDonough and Michael Braungart (2001) have produced a significant criticism of eco-efficiency, however, arguing that the strategy will ultimately fail because it continues to work within the system that initially caused the problem. Eco-effectiveness is an alternative approach that advocates for a "next industrial revolution." The concept describes an industrial system that "is regenerative rather than depletive," and "involves the design of things that celebrate interdependence with other living systems" (McDonough and Braungart 2001, 144). Eco-effectiveness implies that the goal of sustainable production requires a change in our understanding of the nature of the problem and the solution (see Chapter 7).

The realization of a new industrial revolution will require criteria that outline the success factors for sustainable production. These criteria can be applied to three broad sustainability categories of concern: equity or social justice, economy or market viability, and ecology or environmental intelligence (McDonough and Braungart 2001). If new technologies are to be supportable politically, a majority of people in democratic societies must benefit sooner or later from the resulting products or services (Rhodes 2003). To ensure the integrity of the economy, designers must also more closely align industrial innovation with essential social needs. Environmental destruction can be reversed by an emphasis on investing in natural capital and ensuring that new processes do not introduce any hazardous materials into the environment (Hawken et al. 1999). A new design must strive to optimize each of these three components in the sustainability triad as described in McDonough and Braungart's fractal ecology model (McDonough and Braungart 2001). The individual innovators who can combine the various criteria for sustainable production with their design and engineering skills will successfully drive this next industrial revolution (Rhodes 2003).

Innovation is key to ensuring the transition to an eco-effective industrial model. At the level of the firm, innovation can be directly linked to the historical and irreversible process of improvements in production and economic growth derived from technological change (Lipsey and Bekar 1995). Unfortunately, innovation is becoming increasingly difficult as modern society becomes more complex. Institutional organization and numerous barriers that exist across all levels of society are resulting in piecemeal or incremental change, if any change occurs at all (Dale 2001). As a result, innovation in support of sustainable production is most likely to succeed if it results from individuals or the smallest possible units (Rhodes 2003). The role of design is critical to sustainable product innovation in the firm and determines the possibilities set for technologies, materials, and, ultimately, the overall quality of the process or product through its life cycle. In Chapter 4, it has also been shown that firms adopting strategies related to innovation and sustainable production stand to benefit from improved corporate image, cost reduction, further innovation, improved employee morale, and the development of a corporate culture of responsibility.

Interface Flooring, a carpet manufacturer, is a firm that exemplifies all of these characteristics and is renowned as an icon of sustainable production. "A system-wide approach to facility design has enabled Interface to significantly reduce its energy use, to eliminate the use of almost all toxic substances, and to reduce almost all waste from its factories" (Chapter 7). Interface has also transitioned from selling carpet products to leasing floor-covering services so that the firm retains ownership of the product and responsibility for keeping it in good condition (Anderson 1998). The innovations at Interface did not occur through incremental improvements, but rather resulted from a deliberate effort to redesign the flooring business from scratch (Hawken et al. 1999).

Despite the success of Interface in innovative practices, sustainable production, and profitability, it is difficult to cite many other examples of firms that can make comparable claims. Traditionally, there is an array of barriers to innovation within the firm, and innovative insight tends to come from those who are intimately familiar with process rather than those intimately familiar with new technologies (Utterback and Abernathy 1975). The transition of Interface to a service and flow company represents a discontinuous event that occurs rarely in industry if at all and is still not well understood. Firms attempting to innovate often commit substantial resources that yield few positive results, whereas trivial expenditures sometimes produce results of great value. Radical technologies also tend to operate well below their full potential when first introduced, and major structural changes in industrial sectors or society at large are often required before gains are realized (Lipsey and Bekar 1995). Innovation can be successfully managed to overcome barriers, but firms must exhibit more strategic thinking and operate as learning organizations (Dearing 2000; Hawken et al. 1999). Firms and individuals that are able to change design mentalities and adopt a systems perspective toward architecture will improve their chances of successfully innovating and implementing sustainable production practices.

Design, Architecture, and Politics

The critical role of design is a theme that continually re-emerges in the sustainable production literature. Authors frequently discuss the environmental harm caused by the legacy of industrial processes, and cite Einstein's dictum that problems cannot be solved at the level of thinking that created them. In this context, a total "change of design mentality" is required to evade practical limits to innovation through the redefinition of problems (Hawken et al. 1999, 65-66). This proposal is well illustrated by a story that traces the size for the standard US rail gauge back to the spacing of ruts in ancient roads built by the Romans to fit their chariots. "Traditionally poor designs often persist for generations, even centuries, because they're known to work, are convenient, are easily copied and are seldom questioned" (118).

Product design professionals are the crucial interface between users and products, are a linchpin pin of the economy, and are ultimately responsible for the effects of their products on the environment. "The product design and development phase influences more than 80 percent of the economic cost connected with a product, as well as 80 percent of the environmental and social impacts of the product, incurred throughout its whole life-cycle" (Tischner and Charter 2001, 120). As well, the environmental impacts of a product's life cycle are already predetermined after just 1 percent of a project's up-front costs are spent (Hawken et al. 1999). The start of the design process is critically important because for all design problems there may be a variety of routes to a given end rather than a single uniquely determined best practice (Newton and Besley 2002). In order to achieve sustainable production, a different design mentality is required before the process begins.

Ecodesign is one proposed strategy that aims to "integrate environmental considerations into product design and development" (Tischner and Charter 2001, 121). This approach, however, only attempts to incorporate environmental design considerations into conventional product design processes. Sustainable product design is more demanding than ecodesign as it "is concerned with balancing economic, environmental and social aspects in the creation of products and services" (121). In addition, there is currently very little information on the application of sustainable product development, and it is therefore difficult to educate designers in the concept (Tischner and Charter 2001).

The adoption of a systems perspective is recommended to address the problem of producing useful knowledge where sustainable development issues require engagement with complex personal, social, and ecological systems (Dale 2001). "System thinking provides a framework for interrelationships rather than things, for seeing patterns rather than static snapshots. Some of the principles governing natural systems are holism, interdependence and interrelationship" (39). Advocates for a "whole-system" approach to architecture, design, and engineering contend that "optimizing components in isolation tends to pessimize the whole system" (Hawken et al. 1999, 117). Within this design philosophy, a window would not be designed without considering the building, the light without the room, or the motor without the machine that it drives (117).

This theory is further illustrated by the case of Jan Schilham, an Interface engineer, who achieved a twelvefold energy saving for a Shanghai factory by applying the system optimization approach to an industrial process requiring pumps and pipes (Hawken et al. 1999). Schilham optimized the design by simply deploying short, straight pipes and small pumps rather than the traditional design of long and crooked pipes with big pumps that required more horsepower. "That enabled him to exploit their lower friction by making the pumps, motors, inverters and electricals even smaller and cheaper" (116-17). Schilham's innovation is an elegant application of the whole-system approach. The engineer was working on an internal, closed system, however, and the efficiency of the process did not directly impact stakeholders external to the company other than as a production input to the end product. Many firms require the systemic approach to transcend traditional business and political boundaries in order to avoid the undesired impacts that can be associated with "large, interdependent infrastructures" (Dearing 2000, 107).

The transition toward sustainable production raises a number of issues that can easily politicize the design process, thereby delaying innovation and reinforcing the status quo. "Once the potential impact of the product, system or service under consideration has been determined, priorities must be set for further design work. It is rare that all environmental problems can be considered simultaneously" (Tischner 2001, 271). As an additional complication, some possibilities for product improvement may not lie within the firm's sphere of influence. "Knowledge experts in every domain have to realize, accept, and plan for the fact that knowledge of the systems with which they deal is, and always will be, incomplete" (Dale 2001, 39). Consensus regarding future design decisions will continue to become increasingly rare as systems are a constantly moving target and surprise is inevitable (Holling 1998). "This lack of unanimity can be used by competing vested interests as an argument for maintaining the status quo" (Dale 2001, 40). This point is also consistent with the claim that unsustainable designs persist because of a perception that the risk involved in the uncertainty of changing them outweighs any other considerations, such as the impact to the environment (Hawken et al. 1999).

Designer J. Baldwin states that on his first day in design school he was told that "design ... means choosing the least unsatisfactory trade-offs between many desirable but incompatible goals" (Hawken et al. 1999, 112). Baldwin challenges the politicization of the design process, but his assertion is contradicted by a simple and powerful point: "All works of architecture imply a worldview, which means that all architecture is in some deeper sense, political" (Johnson 1997, 44). From the sustainability perspective, a maximum of participation is desirable to address the social and environmental aspects of human development (Spangenberg 2001, 30). This implies that societal institutions should accept more responsibility for governing sustainable production by participating in future planning in order to assist designers in reconciling social, economic, and environmental factors. Unfortunately, these actors are too often constrained by short-term political cycles to accomplish this task effectively. "Business and governments often avoid the task of planning for issues related to the environment or society because the time frames for environmental and social change always seem over the horizon" (Hawken et al. 1999, 316). Ultimately, the future of sustainable production must rest with individuals – the professional MBAs, architects, engineers, and other citizens who create our world (Hawken et al. 1999).

The Changing Role of the Design Professions

An anthropocentric perception of engineers in the twentieth century can be summarized as "the profession of innovators who worked to satisfy a growing desire for products and services and who offered an improved standard of living" (Cruickshank 2003, 24). Nevertheless, the ecocentric ideology that emerged through the latter half of the century more frequently portrays engineers as "ruthless destroyers of the environment whose creations put the health and well-being of the public at risk" (26). Leaders within the profession attempting to address ecocentric concerns propose that engineers involved in product development require "a full understanding of environmental and sustainable development concerns ... as well as the technical and scientific aspects of sustainability issues" before they are able to better act as an interface between society and technological developments (Clarke et al. 2000, 1, 3).

The same advocates are also acutely skeptical about obtaining the commitment to sustainable development from the governments and businesses that would be most capable of driving changes to the profession. "A small group of technical people in positions of corporate and knowledge-based power set the education agenda leaving the rest with a fait accompli" (Clarke et al. 2000, 4). Compounding the problem, engineers with broader backgrounds are often absorbed away from influential technical leadership positions and into managerial roles where their impact on the profession is diluted (Clarke et al. 2000). Design professionals everywhere are also increasingly faced with the challenge of better appreciating the requirement to meet the goals and objectives of projects as determined by clients, government, or wider society: "How the project is achieved is dependent on the local decisions of the engineering team, which is affected by the understanding of the aims and perceptions of the priorities. Underpinning all the decisions are the moral and ethical convictions of the engineer as an individual" (Cruickshank 2003, 26).

Several other intrinsic barriers to sustainable production also regularly confront professional designers in projects: "Compensation paid to architects and engineers is frequently based ... on a percentage of the cost of the building itself or the equipment they specify for it" (Hawken et al. 1999, 91). An engineer's one-time fee looks small compared to the overall cost of the project, but often results in costlier equipment and higher operating costs to the owner. If developers stipulated incentives for achieving efficiency, the problem could be solved. "Perverse incentives for design professionals are only one symptom of a much larger problem" (92), however.

In order to engage a broader set of people in the governance of industrial systems, engineers must open isolated decision-making processes and focus on quality of life rather than specific products. "There is, however, a danger that engineering and scientific methods will be inappropriately applied to an essentially socio-political, cultural and behavioral problem without a full appreciation of the interdependent, chaotic, unpredictable and dynamic nature of the system as a whole" (Clarke et al. 2000, 3). "Properly educated and successful engineers should combine belief, science and political understanding in a unique way, focused on providing practical solutions for real people in various contexts," (Rhodes 2002, 163) but engineering education programs currently lack the focus on politics, art, and psychology required to build this capability into the profession.

No better investments for the future can be made than "improving the quality of designers' 'mindware' – assets that, unlike physical ones, don't depreciate, but, rather, ripen with age and experience" (Hawken et al. 1999, 111). Improving the education of engineers could result in a 20-50 percent increase in equipment efficiency, and over a typical career, a utility would save $6-15 million per brain without taking into account operating costs or pollution (112). Furthermore, a change in product design mentality is irreversible and opens the door to subsequent innovations (Hawken et al. 1999). Engineers should also be encouraged to develop a diverse network of relationships with stakeholders not traditionally associated with the technology development process. These may include environmental activists, ethicists, theologians, ecologists, social activists, government agencies, and concerned members of the public (Clarke et al. 2000, 5). David Orr encourages schools, colleges, and universities to offer both radically different programs of study and a focus on sustainable production issues within their own institutions. Orr believes that changing the procurement, design, and investments made by our educational systems represents a "hidden curriculum" that can teach as "powerfully as any overt curriculum" (Hawken et al. 1999, 315).

A new program at the University of Michigan may serve as a model for graduate students in engineering disciplines to enhance their knowledge of environmental sustainability. The university's College of Engineering recently developed the Concentrations in Environmental Sustainability, or "ConsEnSus," program to expose engineering graduates from all disciplines to environmentally friendlier engineering practices (University of Michigan Engineering Department 2002a). Program director Walter J. Weber Jr. contends: "There is an increasing need for employers to hire new graduates from all disciplines who are informed in matters of environmental regulations, policies, practices and the implementation of clean technologies so that they can anticipate and help circumvent potential problems" (Ritter 2002).

ConsEnSus program developers elected a pedagogical approach that focused on case studies brought to the classroom by real-world engineers. Angela Lueking, a doctoral student who helped launch the program, says that response from the industrial community was better than expected (Ritter 2002). Multidisciplinary teams from Ford, DaimlerChrysler, Dow Chemical, Pfizer, BP, Amoco, and General Electric all participated during the first course offered in the winter of 2002 (University of Michigan Engineering Department 2002b). Michael T. Cannaert, an environmental engineer at Visteon Corporation, one of the world's largest suppliers of parts to the automotive industry, also led one of the case studies and emphasized to students that often there is no clear path to a solution. Cannaert stated: "These complex problems require team-oriented, multidisciplinary approaches. Not only do engineering and science apply, but politics, culture, law and business considerations must be addressed" (Ritter 2002). For sustainable production to be realized, the next phase of its history must bear witness to design professionals receiving more multidisciplinary training, sharing knowledge with a broader set of more non-traditional stakeholders, and taking a central role in explicitly addressing the issues of sustainability in the industrial system.

Ford Model U: You Can Have It Any Colour You Want, as Long as It's Green

The Ford Motor Company also participated in the first class of the University of Michigan's ConsEnSus program by presenting an industrial case study on the sustainable business case for rehabilitation of the firm's Rouge River Manufacturing Complex (University of Michigan Engineering Department

2001). The Rouge River facility was originally the epicentre of the North American industrial revolution and "the most studied and admired manufacturing complex anywhere" (Williams 2002, 61). At one time, the plant encompassed a fully integrated mass production process that employed 90,000 workers and produced everything from Mustangs to airplanes. By the 1980s, however, Ford had spread its manufacturing over 100 sites worldwide and the Rouge River plant became an obsolete environmental catastrophe (Lubell 2003).

Looking for an opportunity to launch the next industrial revolution, eco-effectiveness advocate William McDonough lobbied for a meeting with newly appointed company chair Bill Ford, to sell Ford on his ideas for the future of industrial production. McDonough recalls telling Ford that as head of the giant automakers, Ford could change the world: "With $80 billion of purchase orders, all you have to do is say you want a different way, and things start to move" (Williams 2002, 61). At the end of the meeting, Ford asked McDonough to inspire and lead the remaking of the Rouge River plant into "the model of twenty-first century manufacturing" (Lubell 2003).

In the first phase of a twenty-year, $2 billion overhaul, McDonough, a local architectural firm, and Ford engineers developed a storm water management system at Rouge River to meet anticipated government regulations. The approach utilizes the new factory roof and a contoured parking lot to force runoff through a plant-based filtration system and back to the Rouge River (University of Michigan Engineering Department 2001). The new system is completely natural and cost $35 million less to build than the original designs had forecast for a traditional system (Lubell 2003).

New features at the Rouge River plant also include "skylights to brighten the factory floor and a 'flexible' assembly system" that will enable Ford to "build nine different vehicle models on the same assembly platforms ... potentially saving millions in re-platforming costs" (Lubell 2003). The completion of the first phase of the Rouge River Plant Rehabilitation coincided with Ford's 100th anniversary celebrations in June 2003. The occasion was also used to celebrate another partnership between McDonough and Ford that earlier in the year had resulted in the introduction of the concept design for the world's first recyclable car.

For Ford's centennial, Bill Ford also assembled a design team to create "a Model T for the 21st century" (Ford Motor Company 2003). Ford's vision was to develop a concept car called the Model U that pursued "future automotive technologies without compromising on customer expectations for personal mobility," as inspired by the principles of the Model T (Ford Motor Company 2003). Based on the work McDonough and Braungart had accomplished at the Rouge River plant, Ford asked the pair to be part of a design team for the Model U that also included Ford's Research and Advanced

Engineering, Ford's Brand Imaging Group, BP, and a host of other technology suppliers (Bak 2003).

McDonough and Braungart encouraged the design team to "shift from the 'eco-efficient' strategy ... to pursuing positive environmental effects through intelligent design" (MDBC 2003). They began by placing "the Model U's environmental features and materials into an eco-effective framework, identifying a positive, regenerative vision of what they wanted to accomplish in the long run, and then articulating the ways the Model U begins to move in that direction" (MDBC 2003). McDonough summarizes the philosophy behind the car design, saying: "Model U identifies ways to be recreational and regenerative – to design in a way that is fun and creates environmental benefits at the same time. It offers a totally new vision for the auto industry" (MDBC 2003).

The Model U concept features numerous advanced green materials and processes. The design's centrepiece is a "supercharged hydrogen internal-combustion engine (ICE), which emits only water vapour" (Mateja 2003). Dr. Gerhard Schmidt, Ford Vice-President of Research and Advanced Engineering, believes that "hydrogen will be the automotive fuel of the future ... The hydrogen ICE can act as a stepping stone to hydrogen-fueled mass transportation that will eventually incorporate fuel cells" (Bak 2003). Ford researchers have also shown that with supercharging, "the hydrogen ICE can deliver the same power as its gasoline counterpart and still provide near-zero emissions performance and high fuel economy" (Bak 2003). The Model U can carry 7 kilograms of hydrogen in its on-board fuel tanks, enough to cover a range of about 300 miles. Currently, however, hydrogen is not offered on the mass market as a fuel and must be produced by energy-expending processes further upstream from the consumer. BP currently produces hydrogen from natural gas in limited amounts, but is researching the potential for mass hydrogen production (Jones 2001).

David Wagner, Model U's technology project manager, states that the car design takes a positive approach toward materials and manufacturing and adds that "some of these concepts won't come to fruition for years to come, but this is an important first step" (Ford 2003). Many of the materials in the Model U come from plants, including rubber tires that use corn-based fillers, but only as a partial substitute for carbon black (Ford 2003). There are also several soy-based components, including "the polyurethane seating foam and the tailgate's polyester resin" (Mateja 2003). The Ford research team is working with Shell Global Solutions to develop a lubricant for the Model U based on sunflower seeds, and the company also plans to use eco-effective polyester designed by Milliken and Company in the fabric for Model U seats, dash, steering wheel, headrests, door trim, and armrests (Ford 2003). Rounding out the green material inventory, the Model U's canvas roof and carpet mats are supplied by Interface and are composed of a biopolymer developed by Cargill Dow that is derived from corn and can be safely returned to the soil (Bak 2003).

The Model U also offers a variety of "technical" features considered necessary to market the car in the future. The Model U has the look of an SUV, but has a nearly infinite capacity for being personalized and upgraded, with the ability "to convert the four-door sedan into a four-door pickup, using power controls to retract the top and lower the windows and deck lid" (Ford 2003). The design also offers several technologies to improve driver awareness, including adaptive front lighting, active night vision to enhance images, and cameras mounted on mirrors to enable drivers to see around oncoming traffic at intersections and to prevent and detect collisions.

The Model U is purported to respond to "customers' desires for convenience and entertainment" (Bak 2003). Features include a voice-activated navigation system and an on-board computer that plays music in MP3 format and synchronizes occupant information from personal digital assistants (PDAs) such as address books, personal files, and even driver preference settings for the car (Ford 2003). Bryan Goodman, a Ford technical specialist, suggests that, by taking advantage of the Bluetooth wireless networking technology, electronics can be continually upgraded or added throughout the life of the vehicle (Ford 2003).

While the Model U has been designed for mass production, "its rollout won't be for many years" (Lubell 2003). In addition to the firms developing the green materials for the Model U, Ford is also dependent on suppliers, including Motorola, Sun Microsystems, SpeechWorks, and the Massachusetts Institute of Technology Medialab for the advanced technologies slated for the vehicle. "So far, Detroit insiders have reserved judgment on the worth of the green elements" (Williams 2002, 63). Industry watchers have seen green industrial concepts and radical automotive technologies fail in the past. Paul Einstein, publisher of TheCarConnection.com, asks: "Will these innovations truly work as planned? Will Ford's commitment hold? Will it follow through on other plants, or will this only be the showcase?" (Williams 2002, 63).

Analyst Jim Hall answers: "The truth is, if it's replicable to other industrial sites, it works. If it isn't, it doesn't – but even so, a greener face for the company alone has value, if for no other reason than PR" (Williams 2002, 63). Environmentalists also challenge Ford's motives, asking how the actions of the company will truly be sustaining when the existing production of energy-inefficient SUVs continues to damage the planet and most cities can no longer tolerate growing problems directly attributed to the automotive industry, such as the cost of maintaining roads, traffic congestion, and urban sprawl. McDonough explains: "Yes, Ford makes SUVs ... If you have solar-powered SUVs, which is what we're going to do ... then I don't care, drive around all you want. It's not the problem" (Williams 2002, 63).

Sustainable Production and the Model U

Innovation

The extent to which the Model U is a truly innovative design has yet to be determined. In terms of the criteria for innovation identified in the context of sustainable production, a great deal of uncertainty still surrounds the concept vehicle. The Model U does not meet a unique market need. Many consumers who are currently in the automotive market have a myriad of choices, and there is an increasing menu of vehicles that are becoming environmentally friendly through low-emission engines and hybrid-fuel sources. Many of the technological innovations proposed by the Model U could easily be fitted into existing vehicle designs.

The test-drive is still a watershed sensory metric for car buyers, and Ford would improve its chances to mass-produce the Model U by quickly developing prototypes for consumer tests. In this sense, the Model U would represent an incremental innovation rather than a disruptive innovation, and therefore fewer mistakes should be present in the first vehicles that are produced. Given that the materials in the Model U would be much more resilient and easier to replace, however, it is likely that the life cycle of the vehicle would increase to the point that it would eventually diminish the overall market share of the automotive industry. Ford may be able to gain competitive advantage with the Model U by adopting Hawken and colleagues' service and flow paradigm, as it is unclear how company growth would increase from selling the vehicle as a product.

The Model U concept demonstrates an outstanding vision and application of sustainable production ideas. The design should set an eco-effectiveness standard in the automotive industry and pose a challenge to designers from the other automakers. The effort is also bound to contribute to instilling a culture of environmental responsibility at Ford. It remains to be seen, however, how much of the focus on eco-effectiveness from the Model U design process will live on in the corporate ethos. Most of the inspiration for the Model U is directly attributable to the genius of the McDonough/Braungart team and illustrates how innovation succeeds by occurring in the smallest possible units. Nevertheless, in the long-term, much of the tacit know-how required to mass-produce the vehicle will need to come from the component product and process designers both at Ford and in the rest of the supplier community for the Model U. Those professionals may lack training in eco-effectiveness considerations and not share the same passion for sustainable production solutions as the two architects.

Architecture

From a whole-systems design point of view, the Model U is a tremendous success as it demonstrates the eco-effectiveness benefits of environmental regeneration. There are still numerous political considerations not addressed by the design, however. At this time, it is not clear what the cost of a Model U–inspired vehicle would be. A number of the component technologies have not yet been proven in concept, much less being ready for mass production, and Ford would remain dependent on this broad cross-sector set of companies that lack experience in long-term supply to the automotive industry. Additional scenario planning exercises regarding the mass production and eventual consumer use of the Model U may contribute toward advancing the concept's development within the company. Not factoring cost, the Model U should be acceptable politically as a product, especially in North America, as it fails to challenge the conventional driving lifestyle or look and feel of most mainstream automobiles. Traditional automotive regulations, except for some environmental legislation, would still be required even if all vehicles on the road were Model U's. The use of the Model U still has implications for a number of societal issues regarding the automobile, such as traffic safety, noise and light pollution, and community planning.

It is not known how many non-traditional stakeholders may have been engaged by the Model U team or if any socio-political issues were addressed during the design phase. It is possible that the Model U team had already excessively compromised on the concept before the first day of the design, as corporate baggage would have been brought to the table by the Ford R&D and brand imaging teams. Company research may have previously determined that the look and feel of any concept car would have to remain as the status quo in order to highlight the eco-effectiveness features. Some of the designers may also have had difficulty rationalizing their own ethics and values with those of the Ford Motor Company, McDonough and Braungart, and the other technology suppliers. In the case of the Model U, Bill Ford's corporate leadership is responsible for the advancement of sustainable production in the automobile industry. It may take more concerted efforts by design professionals or political institutions to engage other industrial sectors.

Design Professionals

The concepts of sustainable production, eco-effectiveness, and the Model U suggest that design professionals, a linchpin of the KBE, may be on the frontier of a transition from the ICT revolution at the end of the twentieth century to a new era of eco-innovation. There are, however, considerable technological obstacles that must be overcome before concepts such as the

Model U are fully realized. The transition to a hydrogen economy in the twenty-first century may be the greatest technical challenge ever to confront humankind. Design professionals will need to discover eco-effective processes for the mass production of hydrogen that are not more expensive or more polluting than existing fossil fuels.

As well, the research and development of technologies such as those based on Bluetooth wireless networking dramatically slowed after the 2001 economic downturn in the high-technology sector. The Bluetooth concept has existed since the late 1990s, but few off-the-shelf products implementing the technology have emerged. Design professionals will need to increasingly build bridges between sectors in order to better integrate specific technologies into holistic systems such as the Model U. Ford's concept car also implies significant changes to the automotive maintenance industry. Gas stations could become hydrogen stations and used car dealerships could become upgrade centres. However, the need for eco-innovation, the focus on system architecture, and the resultant redesign of existing support infrastructure required imply substantial changes to existing curricula and retraining efforts for design professionals and technicians across the industry. The University of Michigan's ConsEnSus program is a model that demonstrates the success of engagement between academia, the industrial community, and the design professions on sustainable production issues.

Conclusion

To nurture the next industrial revolution, sustainable production criteria must be employed to assess eco-innovations. This chapter identifies criteria based on innovation, architecture, and the changes to the design professions to establish an assessment framework for sustainable production. It is essential that organizations attempting to achieve more sustainable industrial processes have strong leadership and values that clearly distinguish the commitment of the institution to environmental responsibility. Firms can further encourage innovation by establishing a setting that champions the original ideas of individuals or small teams and allows product design professionals to adopt a holistic approach to system architecture. Organizations can depoliticize product design by governing architecture through multidisciplinary planning and gathering input from a wide range of both traditional and non-traditional stakeholders. Firms must be more proactive in providing feedback to educational institutions and offering training to design professionals in order to build additional capacity in areas such as sustainable production where gaps in curricula currently exist. In order to evaluate the potential of innovations such as the Model U, organizations must employ measurement, simulation, and critical analysis techniques to learn from their experiences and continuously improve overall performance.

Governments can more effectively support the industrial transition to sustainable production by encouraging cross-sector information exchange, ensuring input into architectural development from a broad range of sources, supporting changes to educational programs, and applying all of these practices to their own policy development.

In conclusion, sustainable production is most evident so far in the design of industrial processes rather than products. The cases of Interface Flooring and the Ford Rouge River facility indicate that architects and engineers face fewer barriers to eco-effectiveness under the umbrella of supportive firms. There are still a number of obstacles to the development of sustainable products or services that can replace legacy offerings or present new incentives for customers either through innovation or cost savings. The Ford Model U is an exciting and promising development. If the automotive industry can successfully produce, market, and sell an eco-effective product, the practice should spread quickly to other sectors. If Ford fails to advance the concept further, however, the company will quickly lose credibility and skeptics will once again question the real intentions and commitment of Bill Ford and his designers to the company's environmental responsibilities. Ford carefully managed expectations by not providing cost estimates, a proposed market availability date, or even the next step forward for the Model U after the initial release of the conceptual design. An assessment of the concept against criteria for sustainable production suggests that the design is eco-effective, but cautious toward changing the interface between the product and the consumer. Ford is also dependent on suppliers for unproven technology to compose a new vehicle system that has yet to be prototyped or tested in real-world conditions. Ford should make available to the public its plan for bringing a Model U vehicle to market, with transparent progress indicators that can be monitored and analyzed by interested parties to ensure that all firms are progressing toward making the Model U a reality.

References

Anderson, R. 1998. *Mid-course correction: Toward a sustainable enterprise: The Interface model.* Atlanta: Peregrinzilla Press.

Bak, P.E. 2003. Ford hydrogen powered Model U = Model T + one century. *H2CARSBIZ*, 8 January. http://www.h2cars.biz/artman/publish/article_71.shtml.

Clarke, S., N. Morris, and M. Rhodes. 2000. Managing engineering for a sustainable future. Paper presented at the Challenges for Science and Engineering in the 21st Century conference, Stockholm, June. http://www.inesglobal.org/iee.htm.

Cruickshank, H. 2003. The changing role of engineers. *Engineering Management* (February). http://www.iee.org/Publish/Journals/MagsNews/Mags/Em.cfm.

Dale, A. 2001. *At the edge: Sustainable development in the 21st century.* Vancouver: UBC Press.

Dearing, A. 2000. Sustainable innovation: Drivers and barriers. In *Sustainable development: Innovation and the environment.* Paris: OECD.

Ford Motor Company. 2003. Model U concept: A model for change: Product news. http://media.ford.com/article_display.cfm?article_id=14047.

Hawken, P., A. Lovins, and H. Lovins. 1999. *Natural capitalism: Creating the next industrial revolution.* Boston: Little, Brown.

Holling, C.S. 1998. Two cultures of ecology. *Journal of Conservation Ecology.* http://www.consecol.org/vol2/iss2/art4/.

Johnson, S. 1997. *Interface culture: How new technology transforms the way we communicate.* New York: HarperCollins.

Jones, M.D. 2001. Hydrogen and a lower carbon energy future. Handout. Melbourne: BP Australia. Online: http://www.bp.com.au/news_information/press_releases/hydrogen percent20.pdf.

Lipsey, R., and C. Bekar. 1995. A structuralist view of technical change and economic growth. In *Technology, information and public policy,* ed. Thomas J. Courchene, 9-75. Kingston, ON: McGill-Queen's University Press.

Lubell, E. 2003. Buildings like trees, factories like forests: Ford and the next industrial revolution. *Princeton Independent.* http://princetonindependent.com/issue01.03/item7.html.

McDonough, W., and M. Braungart. 2001. The next industrial revolution. In *Sustainable solutions: Developing products and services for the future,* ed. M. Charter and U. Tischner, 139-50. Sheffield, UK: Greenleaf.

Mateja, J. 2003. Model U taps the cream of the crops. *Los Angeles Times,* 5 February. Online: http://www.latimes.com/classified/automotive/highway1/la-hy-greenford5feb05,0,5649106.story?coll=la-class-autos-highway1.

MDBC. 2003. A model for change: Monthly feature. February. http://www.mbdc.com/News/ModelU_PressRel_20Jan03.pdf.

Newton, K., and J. Besley. 2002. Developing sustainability in the KBE: prospects and potential. Prepared for Carleton Research Unit on Innovation, Science and the Environment (CRUISE).

Presley, A., and L. Meade. 2002. The role of soft systems methodology in planning for sustainable development. *Greener Management International* (37): 101-10.

Rhodes, M. 2002. The world summit points to the need for broader engineering education. *Engineering Science and Education Journal* 11 (5): 162-63. Online: http://ieeexplore.ieee.org/xpl/tocresult.jsp?isNumber=2433.

–. 2003. Opportunity of a lifetime. *Engineering Management* 13 (1): 20-23. Online: http://www.iee.org/Publish/Journals/MagsNews/Mags/Em.cfm.

Ritter, S.K. 2002. Sustainability focus: Michigan engineering graduate programs add study area to enhance environmental literacy. *Chemical and Engineering News* 80 (29): 40. Online: http://pubs.acs.org/cen/education/8029/8029education.html.

Spangenberg, J.H. 2001. Sustainable development: From catchwords to operational benchmarks. In *Sustainable solutions: Developing products and services for the future,* ed. M. Charter and U. Tischner, 24-36. Sheffield, UK: Greenleaf.

Tischner, U. 2001. Tools for ecodesign and sustainable product design. In *Sustainable solutions: Developing products and services for the future,* ed. M. Charter and U. Tischner, 263-80. Sheffield, UK: Greenleaf.

Tischner, U., and M. Charter. 2001. Sustainable product design. *Sustainable solutions: Developing products and services for the future,* ed. M. Charter and U. Tischner, 118-39. Sheffield, UK: Greenleaf.

University of Michigan Engineering Department. 2001. December. http://www.engin.umich.edu/prog/consensus/ford.htm.

–. 2002a. November. http://www.engin.umich.edu/prog/consensus.

–. 2002b. January. http://www.engin.umich.edu/prog/consensus/participants.htm.

Utterback, J.M., and W.J. Abernathy. 1975. A dynamic model of process and product innovation. *OMEGA: The International Journal of Management Science* 3 (6): 639.

Veleva, V., and M. Ellenbecker. 2000. A proposal for measuring business sustainability. *Greener Management International* (31): 90-102.

Williams, F. 2002. Prophet of bloom. *Wired* 10 (2): 60.

World Commission on Environment and Development (WCED). 1987. *Our common future.* New York: Oxford University Press.

Part 3
External and Internal Drivers of Sustainable Production

7

Collaborative Public Policy for Sustainable Production: A Broad Agenda and a Modest Proposal

John Moffet, Stephanie Meyer, and Julie Pezzack

In mid-2000, Shell International initiated an advertising campaign explaining that the oil company was making huge investments in alternative energy sources to avoid becoming "extinct." And for the past four years, Dofasco, Canada's leading steel producer has issued an annual report that addresses the company's "triple bottom line" rather than a traditional financial report. The 2001 report explains that Dofasco measures the success of its performance in the three integrated and mutually dependent components of sustainability – environmental responsibility, social well-being, and financial results (Dofasco 2002).

Shell, Dofasco, and the growing number of other companies similarly reorienting themselves are responding to various important and related dynamics. They are realizing that their contribution to community well-being can directly affect their productivity as well as significantly influence their licence to operate and their image with consumers, business partners, and regulators. They are also recognizing that the environmental pressures of an increasingly crowded and affluent world are starting to threaten the ecological processes that underlie our economic prosperity. Mounting concern over climate change, for example, will eventually induce a carbon-constrained economy in which successful businesses will have to achieve significantly improved energy efficiency or adopt non-carbon–based energy systems. More important for the purpose of this chapter, many companies are making these changes because they see opportunities to improve their corporate image and enhance their market share, productivity, profitability, and competitiveness by adopting new, more environmentally benign modes of doing business.

Together, these dynamics represent the basis of sustainable consumption and production. The premise of this chapter is that sustainable consumption and production represents an important objective for the Canadian economy, not just because it contemplates ecological constraints on some current modes of activity but also because it describes a new way for Canadian businesses to understand and identify how to succeed in the twenty-first century. We argue that this premise has important implications for public policy, which should explicitly and systematically support and encourage the transition to sustainable consumption and production. This chapter observes that a lot of contemporary public policy plays little more than lip service to sustainable consumption and production objectives, and that governments continue to send conflicting policy signals. Finally, it outlines some of the elements of a possible new approach that emphasizes the important synergies between industrial and environmental policies.

Reconceptualizing the Issue

The two approaches currently dominating both environmental policy and the perspective of industrial policies toward environmental issues – environmental protection and eco-efficiency – do not adequately support the objective of sustainable consumption and production. Environmental protection objectives fail to incorporate fully the economic dimensions of the issue. Most importantly, risk-based policies have become the tail that wags the dog. They are so resource-intensive that they force environmental agencies to define their missions very narrowly – away from an analysis of the root causes of unsustainable practices and the development of policies to overcome those barriers.

The response most favoured by industrial policies – eco-efficiency – is an important but incomplete model. First championed by the World Business Council for Sustainable Development in the early 1990s, eco-efficiency essentially exhorts business to "do more with less" (energy, materials, and waste). More efficient production alone is not the answer, however, as it will not necessarily offset the effects of increased consumption. Simply producing the same products more efficiently will not, on its own, ensure sustainable consumption and production. If enhanced eco-efficiency leads to reduced production and product costs, consumption may actually increase. In his famous study of the impact of new technologies on coal consumption in Scotland, Jevons (1865) observed over a hundred years ago that "it is a confusion of ideas to suppose that the economical use of fuel is equivalent to diminished consumption. The very contrary is the truth." More recent studies show that, in many cases, improvements in material efficiency continue to be offset by increasing levels of material consumption (Hettige et al. 1998).

The goal of sustainable consumption and production requires a change in our understanding of the nature of the problem and the solution. In particular, it requires environmental policy to move beyond the current focuses on limiting the direct environmental impacts of production activities, product use, and end-of-life product management to an approach that more explicitly addresses the connections among these activities. McDonough and Braungart (2002) describe this new orientation as "eco-effectiveness." They advocate a model of human industry that is "regenerative rather than depletive" and in which "products work within cradle-to-cradle life cycles rather than cradle-to-grave ones." In the *Harvard Business Review* and their subsequent book, Amory and Hunter Lovins and Paul Hawken describe four interrelated dimensions of "natural capitalism": (1) dramatically increasing the productivity of natural resources, (2) shifting to biologically inspired production models, (3) moving to a solutions-based business model, and (4) reinvesting in natural capital (Lovins et al. 1999; Hawken et al. 1999). More prosaically, the United Nations Environment Programme (UNEP) and the Organisation for Economic Co-operation and Development (OECD) each call for "sustainable consumption and production."

However it is described, this vision entails a combination of stimulating the innovation and dissemination of new technologies and processes and reconceptualizing the way to add value and earn profits in environmentally benign ways. The first component of this vision leads to a policy focus on stimulating innovative ways to design and produce existing products while dramatically reducing environmental impacts throughout the product or service life cycle. This emphasis on innovation is not new to industrial policy, but it does establish a new way to understand the important synergies between industrial and environmental policy.[1]

The second component of this vision is equally important. It focuses on finding new ways to provide services. This requires asking what the function of the service or product is, and how it can be provided most efficiently, taking into account all relevant costs and impacts. It requires asking questions such as whether a utility's customers want energy or the services associated with the provision of energy (lighting, cooling, propulsion, etc.), and whether there may be some other, more efficient means of delivering those services (e.g., through building or process design changes rather than more electricity). In many cases, this focus can be assisted by looking for solutions that are modelled on and respect biological processes, rather than ways that deplete, exploit, or impair them (Benyus 1997). In some cases, this focus also entails redefining the firm as a service provider instead of as a product manufacturer. The resulting focus on function (rather than form) as the source of value to the customer can lead to dramatically changed incentives for environmental stewardship, among other things (Reiskin et al. 2000). Again, the policy focus associated with this direction helps provide a linkage between environmental and industrial policy, a major theme of this chapter.

The oft-cited example of the transition undergone by the carpet company Interface Flooring in the late 1990s illustrates both approaches well.

A system-wide approach to facility design has enabled Interface to significantly reduce its energy use, to eliminate the use of almost all toxic substances, and to reduce almost all waste from its factories. At the same time, Interface moved from selling carpets to leasing "floor-covering services" for a monthly fee, accepting responsibility for keeping the carpet in good condition. This change has been popular with institutional customers who want the benefits of floor coverings without necessarily owning them. It has also created powerful incentives for Interface to design its products so that they will be durable and easily and cheaply maintained, and retain as much value as possible once they are recovered from the customer (i.e., they are easily reusable or recyclable).

Interface's CEO, Ray Anderson, drove that firm's strategy from the top down. The question this chapter addresses is: What public policies are required to induce other companies to follow Mr. Anderson's example, and to do even better?

Reforming Our Public Policies

Market forces drive business decisions. Environmental and other regulations may establish constraints on market activities, but on their own cannot be counted on to promote the kind of innovative behaviour that is at the heart of sustainable consumption and production. To become a widespread, underlying element of our business culture, sustainable consumption and production therefore requires a significant realignment of market forces. At a minimum, this will require:

- mutually coherent environmental and industrial policies focusing on stimulating and disseminating innovation
- new mechanisms and measures for enhancing awareness of opportunities for change on both the demand and the supply sides
- price signals that eliminate inappropriate subsidies and start to incorporate environmental externalities.

This section summarizes four policy measures that could help support these reforms: (1) "smart" regulation, (2) more use of "soft" instruments and partnerships, (3) information disclosure programs, and (4) ecological fiscal reform. The next section then argues that efforts to promote sustainable consumption and production will continue to have limited overall effect until they become fully integrated into industrial policy.

Smart Regulation[2]

Ideally, environmental regulations should address environmental problems while enhancing productivity and stimulating innovation. While there is no easy, one-size-fits-all prescription for smart regulations, there are some basic lessons that have been learned over the past two decades about what works and what does not. In most cases, for example, the most effective approach will be to use a mix of instruments that draws on the strengths of each type of instrument and compensates for its weaknesses (Gunningham and Grabosky 1998).

It is also important to rely on multimedia approaches that give regulated companies the flexibility and incentive to address all their environmental impacts in the manner that results in the most effective overall benefits. Various provinces are now experimenting with "bubble permits," for example. These permits aggregate all of a facility's requirements into a single permit and allow changes in procedures and emissions so long as they do not exceed the caps established under the permit.

In the Canadian context, "smart environmental regulation" requires significantly more federal/provincial/territorial coordination. More explicit linkages between federal regulations and provincial licensing regimes would allow companies to take a whole-facility, multimedia perspective. In our view, a particularly promising way to establish these linkages would be for the federal government to negotiate agreements in which participants commit to specific improvements in overall performance in return for certainty with respect to both federal and provincial interventions. This approach would be similar to the Dutch "Target Group" covenants that have achieved significant success in encouraging cost-effective, innovative solutions to pressing environmental problems.

It is also becoming clear that environmental regulators should focus much more on products, rather than on processes. As Brady and Fava (1999) emphasize, product-focused policies are effective because products are the focal point of many businesses, and addressing products therefore requires a systems perspective that ties environmental issues directly into the core business activities of a company. Brady and Fava argue that this perspective can allow an organization to identify where it can get the best return on investment for its environmental expenditure. It can also stimulate the question at the heart of sustainable consumption and production: what is the core function or service that the product provides, and how can it be provided most profitably and environmentally appropriately?

An emphasis on products should also be accompanied by a perspective that accounts for environmental impacts throughout the life cycle of a product. As Figure 7.1 illustrates, this life cycle perspective can help identify the points of most effective policy leverage.

A key way in which regulations can combine effective solutions to environmental problems with enhanced productivity is by stimulating innovation. Regulators routinely account for the potential economic benefits and costs of different regulatory options. They should also account for potential impacts on technological innovation and seek to stimulate innovation rather

Figure 7.1

Life cycle stages and example objectives and instruments

Life cycle stage	Possible objectives	Possible instruments
Raw material acquisition and processing	*Use low-impact materials* Renewable, non-hazardous, low energy content, recycled/recyclable	Toxics regulations Procurement/supply chain environmental management policies Eco-labels
	Reduce materials Reduce weight and transport volume	Product stewardship/ Extended Product Responsibility (EPR) Pollution prevention (P2) planning
Manufacturing and production	*Clean production* Low, clean energy use Low waste Few, clean inputs	P2 planning Multi-pollutant, facility-wide measures Environmental Management System (EMS) assistance Carbon taxes Waste fees
Filling, packaging, and distribution	*Efficient and clean distribution* Minimize packaging Low-impact packaging Low-impact transport	Packaging content targets, programs, and regulations Supply chain management Waste fees Carbon taxes
Use, reuse, and maintenance	*Minimize user impact* Low, clean energy use Low water/material use and waste Low emissions	Eco-labels Product standards Waste and water fees Carbon taxes
	Optimize initial life Adaptable, upgradable Reliable and durable Easily maintained	
Disposal	*Optimize end of life* Reusable Remanufacturable Recyclable Safely disposable	Disposal fees Deposit/refund Extended Product Responsibility (EPR) Civil liability Waste exchange Eco-industrial parks

than impede it. Too many current environmental regulations favour existing technological solutions. By contrast, regulations could be designed to encourage innovation. They could provide "soft-landing" strategies that do not penalize companies that try innovative approaches but fail and therefore fall out of compliance (Strasser 1996). Regulations could provide exemptions or reduced obligations for companies attempting to develop innovative solutions. Or they could offer long-term certainty as to targets or inspection and reporting obligations for participants in non-regulatory programs designed to seek innovative, "beyond business-as-usual" solutions.[3]

More Use of "Soft Instruments" and Partnerships

Environmental protection regulations will be able to get us only part of the way to a long-term sustainable consumption and production objective. Although an essential part of the mix, environmental protection regulations on their own cannot stimulate the innovation and dynamism required to promote sustainable consumption and production. A wide range of other measures will therefore be required. In particular, we should look to other instruments to stimulate the non-governmental and private sectors to be catalysts for the other, greater part of the way toward our objective.

As Figure 7.2 illustrates, enterprises make environmental decisions in response to a complex and interrelated set of internal and external drivers.

Figure 7.2

Drivers of environmental performance

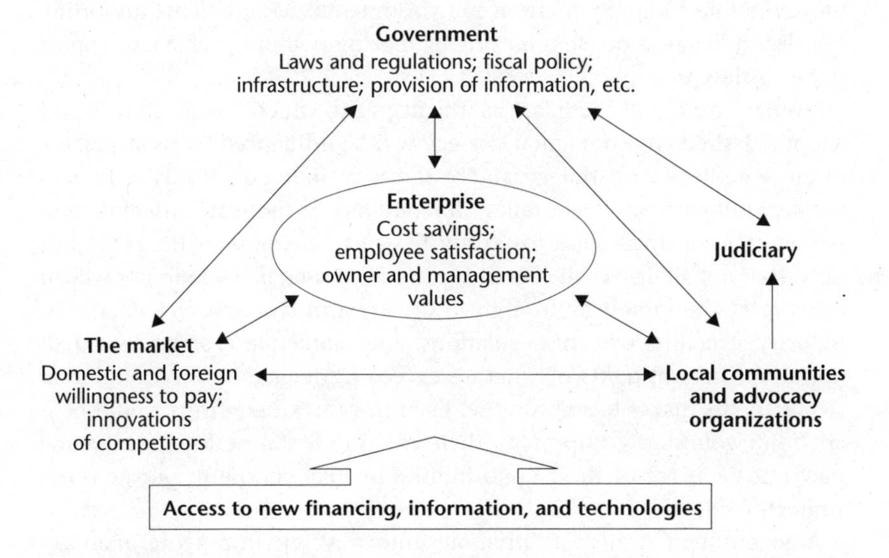

Source: Adapted from Wheeler 1999.

Governments, markets, the local community, and the judiciary all can (and do) create pressure for more or less environmental performance. Governments can use laws and regulations, approval regimes, fiscal incentives, and a wide range of other measures to establish constraints on activities or to create pressure or incentives for desired types of decisions. Well-informed consumers, in turn, can create demand for green products and can reward companies that differentiate themselves from their competitors through green products or high workplace standards (e.g., a commitment to fair wages and no child labour). And, in some cases, emerging market pressure (domestic and international) will compel companies to adopt more eco-efficient measures, either to retain market share or capture new markets. Similarly, local communities can be very effective in pressuring for improved performance.

Each of these drivers can also undermine efforts to promote improved environmental performance. As such, the relative importance of each driver depends on the strength of the others. Where markets do not reward environmental performance and where local communities are complacent, government intervention becomes very important to stimulate market and community pressure (through education, eco-logos, public reporting of releases, information on risks, etc.), to change price signals (resource taxes), or to mandate performance levels (regulation reform and improved enforcement). Conversely, the absence of direct government pressure makes market and community pressures all the more important. Poorly drafted laws and other government instruments can trap the regulated community in a "react-and-control" paradigm. Inappropriate subsidies (such as for energy and water use) or failure to apply or enforce environmental requirements uniformly can distort business decision making by making polluting behaviour appear to be costless.

Overlain on top of these factors, the degree to which any given firm will adopt a desired environmental strategy will be influenced by its awareness of the benefits of that strategy, and by the accessibility of affordable financing, technologies, and techniques. Different factors motivate different companies. The business case for going beyond "business as usual" is not universal: not all firms will benefit equally from such behaviour, nor will all firms react the same way to a given driver. Firm size and characteristics, industry structure, existing regulations, and marketplace pressures are all important determinants of whether exceeding regulatory environmental requirements makes business sense. Even in cases where firms would benefit from voluntarily improving their environmental performance, some may not be aware of these opportunities or may choose to pursue other strategies.

A government strategy to promote innovative environmental management thus needs to account for:

- the degree to which each of the above drivers currently creates incentives for innovation
- the relative importance of existing barriers
- the likelihood that any existing barriers can be removed
- the anticipated benefit of strengthening each category of driver
- the likelihood of being able to strengthen each category.

Government policy to promote innovative environmental performance must also account for the fact that governments can influence some but not all of the business drivers for beyond business-as-usual actions. The purpose of government incentives is to enhance the effect of existing business drivers. As Figure 7.3 illustrates, however, governments do not influence all of these drivers equally. For example, although governments have a lot of leverage in reducing firms' regulatory burden and they can influence a firm's public image, they are unlikely to persuade a firm to change its internal values.

Based on the above insights, we suggest that government policies to promote innovative environmental performance may need to be designed differently from policies focusing on ensuring compliance with specific environmental protection objectives. Such policies may need to be more open-ended. They may need to focus more on incentives than penalties.

Figure 7.3

Government leverage over environmental management drivers

	Departments					
	Environment				Finance	All
Incentives/ drivers	Regulatory relief	Public disclosure	Recognition	Technical assistance	Tax incentives	Linkage to other programs
Reducing regulatory costs	✓✓					
Reducing costs				✓	✓✓	✓
Gaining competitive advantage	✓			✓	✓	
Protecting or increasing sales			✓			✓
Enhancing public image		✓	✓			
Reflecting internal values						

And, in many cases, they may be best provided by departments other than environmental protection agencies.

The Canadian experience with policies supportive of such actions is evolving slowly, and may be somewhat underdeveloped relative to the international experience. Many European countries, almost half of the states in the US, and a growing number of developing countries have introduced flexible programs and incentives focusing on promoting "leadership," "beyond compliance," or "beyond business-as-usual" changes. In Canada, there is considerable experience with partnership activities and other types of non-regulated or extra-regulatory initiatives focusing on habitat conservation. There is little Canadian experience, however, with the provision of incentives to alter polluting behaviour, such as regulatory relief (related to permits and regulatory certainty) or public disclosure programs.

A report on reform options prepared for the Ontario Ministry of Environment (Executive Resource Group 2001) summarized the current situation well:

> Despite internal awareness of these approaches, the Ministry to date has not made progress towards articulating this vision more fully and developing the political and public consensus, including policy, program, and organizational options, to make it a reality. It was apparent to us that a core of the Ministry is firmly entrenched – philosophically, culturally, and programmatically – in a traditional command and control approach. While there are examples of leading edge-type initiatives emerging from various creative centres in the organization, these do not fundamentally challenge the traditional approach.

The most important Canadian example is probably Environment Canada's Accelerated Reduction/Elimination of Toxics (ARET) program. With limited government funding, ARET helped raise the profile of toxics and pollution prevention within many industrial sectors, and helped contribute to sizable reductions by some participants.

Based on the lessons learned from ARET and a handful of similar provincial challenge programs, the federal government and a few provinces have recently attempted to develop new, more rigorous challenge programs, primarily based on some type of formal agreement between government and participant. It is to be hoped that the agencies in question provide the support and incentives required to ensure that these programs are effective.

Admittedly, the international experience with these types of measures is still relatively recent and, partly as a result, the effectiveness of such policies remains largely unevaluated (OECD 2003). Nonetheless, in our opinion, there is sufficient anecdotal evidence to suggest that increased use of these approaches would represent an important addition to the Canadian policy maker's toolkit.

To summarize the argument thus far, environmental regulations are only part of the solution, and government should deploy programs that create incentives for beyond business-as-usual performance. The design of such programs will not be easy. Businesses make environmental decisions in response to a complex and interrelated set of internal and external drivers, only some of which governments can influence. In addition, each driver can also undermine efforts to improve environmental performance, and the business case for going beyond business-as-usual is not universal. Different businesses will respond differently to the wide range of regulatory, economic, and social processes they each face. As a result, a strategy to stimulate innovative, beyond business-as-usual action must deploy an array of incentives, including many that rely on non-regulatory and even non-governmental actors and forces.

Information Disclosure Programs

Information disclosure programs represent one kind of tool that can give considerable power to non-governmental actors to create pressure for improved environmental performance. An important study of the US Toxics Release Inventory (TRI) concluded that the most important impacts of the TRI have flowed from its use by third parties to target the most egregious polluters for direct community action ("environmental blacklisting") and to pressure government to focus on tightening standards and enforcing existing ones against those facilities at the bottom of the list (Fung and O'Rourke 2000). The authors characterize this effect of focusing maximum public attention on the worst (minimum) performers as "Populist Maxi-Min Regulation." They argue that this approach differs from traditional command and control in that the role of public agencies is not to set and enforce standards but to establish an information-rich context for private citizens, interest groups, and firms to solve environmental problems. In this context, standards are set at levels informed citizens and the market will accept; corporate environmental management decisions are made in response to a range of public pressures rather than to formalized agency standards or government sanction; and public pressure "ruthlessly focuses on the worst polluters – maximum attention to minimum performers – to induce them to adopt more effective practices."

Fung and O'Rourke (2000) argue that this focus on linking mandated public disclosure with support for non-governmental actors can be replicated in other contexts at a fairly low cost to government. At least twenty-six US states now have "generation disclosure" rules requiring electricity providers to disclose to customers the type of energy used to generate the electricity being supplied. Similarly, a growing body of experience from the US and elsewhere points to the positive impact on innovation of other information-based instruments such as eco-labels, eco-audits, and liability regimes (Kemp 1997; Cleff and Rennings 1999).

In this regard, we think that governments have an important role to play in encouraging the investment markets to account directly for environmental performance. The growth in demand for socially responsible and "green" investment opportunities has been impressive over the last few years. Investment in these areas in Canada is growing at 40 percent per year, and now includes $50 billion, or 4 percent of total assets. In the US, the numbers are higher: $2.2 trillion, equalling 13 percent of total assets under management (Social Investment Organization 2003).

Notwithstanding this growth, however, recent work by the World Resources Institute indicates that mainstream investment markets are unlikely to routinely account for the full dimensions of environmental performance until investment analysts are able to understand and measure more easily the links between environmental and financial performance. Recent initiatives in some of the Scandinavian countries and Great Britain are illustrating that governments – industry and finance departments in particular – can play a very important role in supporting this transition. For example, governments could require more complete corporate disclosure and could also play active roles in helping the investment community develop and learn how to use new analytical techniques that would make these considerations part of routine financial analysis.

While both environmental and industrial policy agencies can encourage and help companies to identify possible win/win opportunities resulting from improved environmental performance, the permissible scope of non-financial issues that managers and boards of directors of publicly traded companies are legally allowed to address is typically well outside the domain of environmental agencies. Although polls indicate that most Canadians want companies to address a wide range of social and environmental issues, current corporate governance laws in Canada are unclear as to the degree to which this would be permitted, and most conventional interpretations restrict corporate managers and boards to considering *only* issues with possible direct impacts on the value of the company (this is their "fiduciary duty"). Many commentators, including most recently the well-publicized Canadian Democracy and Corporate Accountability Commission (2002), have recommended expanding this duty to enable managers and boards to consider non-financial considerations related to the environment and corporate social responsibility. This is an issue that may be more applicable to North American companies than others, and is certainly gaining prominence on our side of the Atlantic (see Chapter 9).

Ultimately – Ecological Fiscal Reform

In order to change consumption and production patterns fundamentally, it will be essential to integrate environmental costs into product prices through the tax system. Economists estimate the value of the earth's ecosystem services to be at least $33 trillion a year (Costanza et al. 1997). Yet, because the value of these services does not appear on any balance sheet, companies account at most for their use of resources, not for their impacts on ecosystem functions. Although we cannot expect individual businesses to voluntarily account for costs that are shared by everyone, our tax system can and should start to incorporate these costs. But it does not. Indeed, in some cases, the tax system actually creates a preference for environmentally harmful products or activities.[4] As OECD Secretary-General Donald Johnston stated: "Managing resources so as to support sustainable development ... requires internalizing negative production and consumption externalities, for example, through reforms of subsidies that are harmful to the environment, the use of economic instruments such as taxes and charges, the creation of markets, and better appraisal of external effects" (Johnston 2000).

As we define it,[5] ecological fiscal reform (EFR) has five key elements:

- removal of existing incentives and subsidies to environmentally unsound practices
- removal of existing fiscal disincentives to environmentally sound practices
- use of eco-taxes to help internalize the true costs of production and consumption
- selective use of incentives to encourage desired behaviours (e.g., accelerated capital cost allowances for energy- and material-efficient technologies, revenue-neutral taxes on products and practices with high externalities, etc.)
- development and use of new measures of progress that account more fully for environmental impacts than the current system of national accounts.

Growing evidence from Europe, in particular, suggests that EFR could be introduced in a revenue-neutral manner that produces net benefits for the economy (e.g., by using revenues from eco-taxes to reduce employment and other taxes that actually deter economic growth) (Moffet et al. 2003).

Despite considerable rhetoric and interest in this area, Canada has made disappointing progress. Remarking on the limited ecological fiscal reform activity in Canada, the National Round Table on the Environment and the Economy (NRTEE) initiated a review of opportunities for EFR in 2000. In a 2002 overview report, the NRTEE concluded that Canada's limited use of economic instruments has generally followed the US model of market-based approaches targeted toward environmental protection, rather than the revenue-neutral, double-dividend tax-shifting approach that is gaining strength in Europe.

Following the poorly publicized Task Force on Economic Incentives and Disincentives to Sound Environmental Practice (1994), the federal government has had some success at removing the most egregious subsidies to environmentally damaging activities. In particular, the program review exercise in the mid-1990s led to the removal of many agricultural and energy megaproject subsidies with perverse environmental effects, although this exercise was driven by deficit reduction rather than environmental objectives.

At the federal government level, there have been a few modest experiments with tradable permits. The federal government has used a tradable allowance system to eliminate methyl bromide (an ozone-depleting substance) and has provided limited support to two pilot emissions-reduction trading schemes. There have also been a number of environmentally inspired tax incentives, including differentiated excise taxes on leaded and unleaded gasoline, excise tax exemptions for alternative fuels such as ethanol, measures to level the playing field between conventional energy and renewable energy sources, and tax benefits for gifts of ecologically significant land. Environmentally focused program expenditures and subsidies have been more common, beginning with the $3 billion Green Plan in 1990, green procurement policies, grants and loans for research and development of various sustainable technologies, and a range of funds to support education and action programs. The federal government has also continued to provide various fairly modest incentives for energy efficiency and renewable energy (Government of Canada 2005; VanNijnatten 2002).

The use of economic instruments is more common at the provincial and municipal levels in Canada.[6] Ontario has launched an emissions credit trading system for smog precursors, and some municipalities provide tax breaks to help conservation objectives. The most ambitious effort to date has been the British Columbia Green Economy Initiative.[7] That initiative supported a pilot project that combined permit fees and rebates for investments in alternative technologies to help phase out the sawmill industry's remaining beehive burners and unmodified silo burners in advance of a regulatory phaseout scheduled for the end of 2004. Unfortunately, however, the Green Economy Secretariat responsible for this initiative was disbanded in September 2001, and BC has not pursued further tax-shifting initiatives.

Many provincial governments also use economic instruments for product stewardship. Examples include deposit-refund schemes for beverage containers and advance disposal fees for tires and used oil. In most cases, however, the revenue from these instruments is not recycled for environmental purposes, contributing to public skepticism and resistance to the use of economic instruments for environmental protection.

Notwithstanding the above examples, Canadian experience with EFR has been limited (Moffet et al. 2003). The 2000 OECD economic survey of Canada evaluated Canadian policies for sustainable development, and noted: "No-cost opportunities for curbing pollution are rare, and a strategy based on voluntary agreements alone cannot be expected to correct completely for the external costs of pollution. Hence there is a need to increase the use of economic instruments (for instance, charges on toxic emissions and waste, and disposal fees for products containing toxic substances) to reinforce the polluter-pays principle" (OECD 2000, 7).

Although it has not embraced a comprehensive ecological fiscal reform agenda, Canada has tentatively opened the door to important – and unique – progress with respect to the development of new indicators of national progress. Observing that "we must come to grips with the fact that the current means of measuring progress are inadequate," the minister of finance announced the Environment and Sustainable Development Indicators (ESDI) Initiative in the 2000 spring budget. He stated that the indicators developed by this initiative "could well have a greater impact on public policy than any other single measure we might introduce." The ESDI Initiative's report (NRTEE 2003) included detailed recommendations for expanding the current *System of National Accounts* to include broader measures of the human, natural, and social capital on which current and future development depends. The initiative also recommended the annual publication, preferably with the federal budget statement, of a small set of national indicators of some of the key aspects of natural and human capital. Adoption of the ESDI Initiative's recommendations would place Canada among the forefront of countries striving to promote increased awareness of the environmental and social bases of our well-being and of the full consequences for future generations of our current activities.

Aligning Environmental and Industrial Policy

A recent review of governance models for environmental protection observed that "in many jurisdictions, the primary responsibility for the environment has been delegated for the most part to one department of government. Leading jurisdictions are recognizing that the challenge of effective environmental management is broader than one department. There is a growing awareness that the solutions can only be achieved by marshalling and aligning all of the resources of government to achieve a common purpose" (Executive Resource Group 2001).

This conclusion echoes the views of many others. A recent review of environmental governance in Europe, for example, observed that "the problem faced by environmental departments the world over is that the driving forces of pollution, habitat loss and development reside in virtually every sectoral ministry of the state. Indeed, with the exception, perhaps, of economic policy, environment is probably the most cross-cutting of all policy issues confronting government" (Jordan 2002, 38). Former European Union environment commissioner Ritt Bjerregard made the same point when she observed that she was "a bit like someone in charge of a car park where none of the issues which are parked there under the name of the environment are really ones that I could call my own. In reality, they are in fact issues which really need to be resolved elsewhere by some of my other colleagues, with responsibility for agriculture, industry, and energy" (Bjerregard 1995).

One of the key tenets of the argument in this chapter is that an innovation-based policy requires close integration and alignment of industrial and environmental policies. As Figure 7.3 illustrates, environmental agencies do not control the important government levers. Industrial policy agencies (including agricultural, transport, resource development, and other industry-focused agencies) should see themselves as key partners with environmental agencies in promoting the kind of change called for in this chapter. At a minimum, these agencies should actively promote the vision advocated here of an economy that seeks out the potential synergies among environmental and economic objectives.

More fundamentally, it may be worth considering a more profound reordering of responsibilities. Perhaps environmental agencies should focus on their core mandate of environmental protection. Perhaps they should conduct science, set limits, and monitor environmental quality and compliance levels. And perhaps we should task other government agencies with the responsibility to promote innovation-based, beyond business-as-usual environmental performance.

Although this recommendation may be seen as a radical departure from established government agency orientation, it may be worth considering for two reasons. First, environmental agencies are unlikely to be effective in promoting the kinds of policies advocated in this chapter and elsewhere. The more their activities resemble or overlap with industrial policy, the less effective environmental agencies are likely to be. Environmental agencies have neither the institutional mandate nor the internal expertise to deliver such programs, and attempts to do so will be resisted by those agencies whose mandate and expertise are being encroached upon. Second, and conversely, the more the policy measures required overlap with traditional industrial policy considerations and activities, the more effective industrial agencies are likely to be in delivering them.

Regardless of whether this sort of realignment occurs, governments will be successful at promoting sustainable consumption and production only if industrial policy agencies become much more involved in developing and deploying levers to promote innovation-based, beyond business-as-usual environmental and economic performance. Such levers could include support for the development, improvement, and dissemination of:

- technologies that are consistent with the vision of "eco-effectiveness" or "natural capitalism"
- best management practices (such as environmental management systems, supply chain management, pollution prevention planning, and product stewardship)
- standardized analytical tools (such as life cycle analysis)
- standardized analytical measures (such as life cycle–based eco-efficiency indicators)
- accurate consumer information (such as product standards, product labels, and meaningful, comparable corporate sustainability reports).

Analytical tools, measures, and public information form a particularly important bundle. Emerging experience such as the US Green Carrot Program also emphasizes the importance of carefully structured government/industry partnerships in stimulating "radical" innovations. Together, these types of approaches are the key to creating the sort of competitive dynamic within our marketplace that will be required to move us beyond where even the "smartest" environmental regulations will take us.

Although the Canadian experience with this type of integrated approach is limited almost exclusively to the promotion of energy efficiency, other countries have gone much further toward integrating environmental and industrial policies. The United Kingdom merged its departments of transport and environment. Dutch departments of industry and environment share responsibility for the overarching National Environmental Plan goals. And a recent review of Danish policies observed that "key economic departments, such as trade and industry, are now beginning to see environment policies on climate change, for example, as a positive enabler of wealth generation, rather than a crude barrier to innovation and enterprise" (Lenschow 2002, 40).

Corporate Sustainability Reporting: An Example of the Desired Policy Focus and Partnership Approach

Interdepartmental partnerships offer one mechanism by which governments can better align environmental and industrial policy objectives. By jointly signalling support for a common policy objective and using a range of levers to encourage action toward this objective, departments can increase their individual influence over the targeted community. A Canadian example of this approach is the joint work by three federal departments – Environment Canada, Industry Canada, and Natural Resources Canada – to promote enhanced voluntary corporate disclosure of sustainability performance information through corporate sustainability reports.

The policy question these departments have each addressed is: "Why should governments promote corporate sustainability reporting?" The benefits of increased and improved sustainability reporting can include enhanced awareness and corresponding action both within the reporting company and externally. Internally, the process of preparing a corporate sustainability report enables companies to document their own resource inputs and non-product outputs. This can help them to identify previously overlooked areas requiring improvement. "Reporting itself is an amazing driver for change within organizations. They commit to saying something without necessarily knowing what that commitment means, and then they go through the process of trying to gather the data. That process alone causes the penny to drop in many organizations" (Lines 2001). Documenting this information is also necessary in order to implement pollution prevention or cleaner production initiatives.

Externally, these reports provide a range of stakeholders and decision makers – including shareholders, financial institutions, investors, customers, and communities – with information they need to make informed decisions, allowing them to reward companies with strong sustainability performance and punish companies with a weak performance record. Encouraging companies to produce high-quality sustainability reports constitutes one element of a broad information disclosure program.

These reports can also provide governments with a valuable source of corporate environmental and social information, furthering their understanding of current management practices and current consumption and emission practices for specific inputs and substances. This information can support informed decisions on how best to achieve specific government objectives. Furthermore, as governments continue to encourage beyond business-as-usual behaviour through participation in voluntary programs, the primary way to monitor the effectiveness of these programs and to enhance their public credibility is through the public reporting of progress.

Governments can encourage voluntary sustainability reporting in various ways:

- They can create public interest and demand by raising awareness about the role and proliferation of corporate sustainability reports.[8]
- They can provide tools, such as reporting guides, or support the development of standardized reporting formats and indicators and measures.
- They can provide incentives to companies that prepare reports meeting specified standards.[9]

- They can facilitate access to sustainability reports through a centralized report clearinghouse.[10]
- They can make reporting a participation requirement of government-led voluntary programs.
- They can require such reports through statutory or regulatory provisions.[11]

The Canadian government began its promotion of corporate sustainability reporting through sponsorship of a detailed benchmark survey of corporate environmental and sustainability reporting in Canada to gain a better understanding of current reporting practices and to identify opportunities for improvement.[12] This survey, led by Stratos Inc., was an example of the kind of collaborative activity advocated in this chapter, bringing together an active sponsor group that included three federal and one provincial government departments, leading corporate reporters, and representatives of the investment community. On their own, each of these partners can help promote more widespread, good-quality reporting. The collaboration and exchange of ideas fostered by interaction among the sponsors encouraged each to identify additional ways to support increased and improved reporting. In addition to promoting significant changes in corporate reporting practices in some companies included in the survey, the process provided considerable stimulus to government efforts in this area – stimulus that might not have arisen without its participation in such a multi-stakeholder process.

The federal departments continue their collaborative work in this area. In an effort to understand the barriers to voluntary corporate disclosure of sustainability information, Environment Canada, Industry Canada, and Natural Resources Canada have consulted with industry associations and their members to determine the reporting barriers companies face and to solicit input on the possible role the federal government could play to lower some of these barriers in order to improve the quantity and quality of Canadian reporting practices. Based on this understanding of barriers, the federal departments produced a sustainability reporting communications package designed to educate companies about sustainability reporting and its benefits. In addition, Environment Canada, Industry Canada, and the Department of Foreign Affairs and International Trade have co-sponsored a reporting "toolkit" that addresses the information needs identified by industry (SustainAbility Ltd. and UNEP 2000).

These collaborative efforts, both across government departments and in partnership with industry, send consistent signals to producers and users of corporate sustainability reports that the federal government promotes voluntary corporate disclosure of sustainability information as one important way to encourage improved environmental and sustainability performance. It is unlikely that any single department – and almost certainly not Environment Canada acting on its own – could have been as effective.

Conclusion

Although a select few Canadian companies are at the forefront of the transition to a more sustainable mode of production, Canadian public policy has not yet fully embraced the kinds of changes required to ensure that our economy takes advantage of this transition. We recognize that the kinds of changes called for in this chapter will not be easy to make. Our economy remains heavily reliant on resource industries for which some aspects of this model may appear threatening. Moreover, any significant change in policy orientation will have to overcome concerns about departing too far from the dominant approaches in the United States, our largest trading partner.

Implementing the types of policies needed will also require major changes for Canadian governments. In particular, adopting the recommendations in this chapter would lead to more:

- interdepartmental, interjurisdictional, and public/private collaboration
- focus on product design and consumption drivers rather than on environmental impacts
- use of non-regulatory measures
- focus on indirect, information-based levers such as standardized analytical tools and measures, public access to information, and corporate environmental and sustainability reporting
- emphasis on fostering corporate leadership
- focus on education and training
- use of appropriate fiscal measures.

The overall vision articulated here is a challenging one. It suggests that there are choices that can address many of the root causes of our current environmental issues while enhancing our prosperity. This is not to imply, however, that implementing this vision will be easy or that it will not require significant and difficult trade-offs. Nor should one view the specific policy reforms proposed here as a comprehensive solution. We present them as examples of the desired direction and type of policy measures required to support the transformation we believe is required. That said, failure to undertake these changes could lead to unnecessary environmental damage and missed economic opportunities.

Notes

1 This focus does not assume that all environmental impacts can be reduced or eliminated without cost: significant trade-offs may be required on a case-by-case basis.

2 We first prepared this chapter in 2001, before former prime minister Jean Chrétien established the Advisory Committee on Smart Regulation. The views expressed in the chapter are similar to those in the Advisory Committee's report (EACSR 2004), but focus on environmental regulation specifically.

3 For a longer discussion of options for statutory incentives for participating in non-regulatory measures, see Moffet et al. 2002.
4 See, for example, the sections of *Final Report* of the Task Force on Economic Incentives and Disincentives to Sound Environmental Practice (1994) addressing the tax treatment of virgin versus recycled materials.
5 See Moffet et al. 2003.
6 For a partial listing and description of these, see Barg et al. 2000.
7 For more information on the BC Green Economy Initiative, see Baker 2001.
8 In a speech made to the Business Action for Sustainable Development (BASD) in Johannesburg, South Africa, during the World Summit on Sustainable Development, 1 September 2002, Canada's prime minister challenged companies to report on the environmental and social impacts of their operations.
9 For example, the Wisconsin Department of Natural Resources (2001) recently signed an environmental cooperative agreement with Wisconsin Electric that, in exchange for certain benefits (including permit streamlining, alternative monitoring, and more flexible operations), requires the company to meet certain requirements, including preparation of an annual environmental performance report in accordance with the Global Reporting Initiative (GRI) *Guidelines* (2000). This is the first known specification of the GRI *Guidelines* in a legal agreement.
10 For example, Environment Australia operates a clearinghouse for electronically available public environmental reports, with links to company sites, publications, tools, and other resources: http://www.ea.gov.au/industry/sustainable/per. The Australian government is currently supporting an initiative to post standard information about corporate environmental and social performance on the Web to facilitate comparability of information, primarily by fund managers and investors (see http://www.sirisdata.com).
11 Denmark and France have mandated environmental and social reporting. France, for example, requires companies listed on the stock exchange to report information on the environmental and social impact of their activities – such as greenhouse gas emissions and effects on biodiversity – in their annual reports.
12 The Canadian benchmark survey results are presented in *Stepping Forward: Corporate Sustainability Reporting in Canada* (Stratos 2001) and its successor, *Building Confidence* (Stratos 2003).

References

Baker, K. 2001. Tax shifting in BC. Presentation to the National Round Table on the Environment and the Economy Ecological Fiscal Reform Program Expert Advisory Group. Victoria: Green Economy Secretariat.

Barg, S., et al. 2000. *Analysis of ecological fiscal reform activity in Canada.* Winnipeg: International Institute for Sustainable Development.

Benyus, J. 1997. *Biomimicry: Innovation inspired by nature.* New York: William Morrow.

Bjerregard, R. 1995. Speech to the Royal Society of Arts. London: Green Alliance/ERM.

Brady, K., and J. Fava. 1999. *Product and supply chain focused policies and tools for sustainable development.* Ottawa: Five Winds International.

Canadian Democracy and Corporate Accountability Commission. 2002. *The new balance sheet: Corporate profits and responsibility in the 21st century.* Toronto: CDCAC.

Cleff, T., and K. Rennings. 1999. Determinants of environmental product and process innovation. *European Environment* 9: 191-201.

Costanza, R., et al. 1997. The value of the world's ecosystem services and natural capital. *Nature* 387: 253-60.

Dofasco Inc. 2002. *2001 Annual Report.* Hamilton, ON: Dofasco. Online: http://www.dofasco.ca.

Executive Resource Group. 2001. *Managing the environment: A review of best practices.* Report prepared for the Ontario Ministry of Environment.

External Advisory Committee on Smart Regulation (EACSR). 2004. *Smart regulation: A regulatory strategy for Canada.* Ottawa: EACSR. Online: http://www.pco-pcb.gc.ca/smartreg-regint.

Fung, A., and D. O'Rourke. 2000. Reinventing environmental regulation from the grassroots up: Explaining and expanding on the success of the Toxics Release Inventory. *Environmental Management* 25 (2): 115-27.
Global Reporting Initiative (GRI). 2000. *Sustainability reporting guidelines on economic, environmental and social performance.* Boston: GRI.
Government of Canada, Department of Finance. 2005. *Budget 2005: Delivering on commitments.* Ottawa. Online: http://www.fin.gc.ca/budtoce/2005/budliste.htm.
Gunningham, N., and P. Grabosky. 1998. *Smart regulation: Designing environmental policy.* Oxford: Oxford University Press.
Hawken, P., A. Lovins, and H. Lovins. 1999. *Natural capitalism: Creating the next industrial revolution.* Boston: Little, Brown.
Hettige, H., M. Mani, and D. Wheeler. 1998. Industrial pollution in economic development: Kuznets revisited. World Bank Policy Research. Working Paper No. 1876. Washington, DC: World Bank. Online: http://www.wds.worldbank.org.
Jevons, W. 1865. *The coal question.* Reprint of the 3rd ed. (1906). New York: Augustus M. Kelly.
Johnston, D. 2000. Sustainable development – An OECD perspective. *International Council on Metals and the Environment Newsletter* 8.
Jordan, A. 2002. Efficient hardware and light green software: Environmental policy integration in the UK. In *Environmental policy integration: Greening sectoral policies in Europe,* ed. A. Lenschow. London: Earthscan.
Kemp, R. 1997. *Environmental policy and technical change: A comparison of the technological impact of policy instruments.* Cheltenham, UK: Edward Elgar.
Lenschow, A. ed. 2002. *Environmental policy integration: Greening sectoral policies in Europe.* London: Earthscan.
Lines, M. 2001. The 2001 benchmark survey. *Business and the Environment* (May).
Lovins, A., H. Lovins, and P. Hawken. 1999. A road map for natural capitalism. *Harvard Business Review* (May-June): 145-58.
McDonough, W. and M. Braungart. 2002. *Cradle to cradle: Remaking the way we make things.* New York: North Point Press.
Moffet, J., W. Davis, and B. Mausberg. 2002. *Supporting negotiated environmental agreements with statutory and regulatory provisions: An overview for Ontario.* Toronto: Environment Defense Canada.
Moffet, J., F. Bregha, and M.J. Middelkoop. 2003. *Economic instruments for environmental protection and conservation: Lessons for Canada.* Prepared for the Government of Canada's External Advisory Committee on Smart Regulation. http://www.pco-bcp.gc.ca/smartreg-regint.
National Round Table on the Environment and the Economy (NRTEE). 2002. *Toward a Canadian agenda for ecological fiscal reform: First steps.* Ottawa: NRTEE.
–. 2003. *Environment and sustainable development indicators for Canada.* Ottawa: NRTEE.
Organisation for Economic Co-operation and Development (OECD). 2000. *Economic surveys: Canada.* Paris: OECD.
–. 2003. *Voluntary approaches for environmental policy: Effectiveness, efficiency and usage.* Paris: OECD.
Reiskin, E., J. Kauffman Johnson, and T. Votta. 2000. Servicing the chemical supply chain. *Journal of Industrial Ecology* 3 (2): 19-31.
Social Investment Organization (SIO). 2003. *Canadian social investment review 2002.* Toronto: SIO.
Strasser, K. 1996. Preventing pollution. *Fordham Environmental Law Journal* 8 (1): 1-58.
Stratos Inc. 2001. *Stepping forward: Corporate sustainability reporting in Canada.* Ottawa: Stratos. Online: http://www.stratos-sts.com.
–. 2003. *Building confidence: Corporate sustainability reporting in Canada.* Ottawa: Stratos. Online: http://www.stratos-sts.com.
SustainAbility Ltd. and United Nations Environment Programme (UNEP). 2000. *The global reporters.* London: SustainAbility.

Task Force on Economic Incentives and Disincentives to Sound Environmental Practice. 1994. *Final Report.* Ottawa: Department of Finance.
VanNijnatten, D. 2002. Getting greener in the third mandate? Renewable energy, innovation and the Liberal's sustainable development agenda. In *How Ottawa spends: 2002-3003,* ed. G.B. Doern, 216-32. Toronto: Oxford University Press.
Wheeler, D. 1999. *Greening industry: New roles for communities, markets and governments.* Washington, DC: World Bank.
Wisconsin Department of Natural Resources. 2001. *Environmental cooperative agreement between Wisconsin Electric Power Company and Wisconsin Department of Natural Resources.* Madison: Wisconsin Department of Natural Resources.

8
Mobilizing Producers toward Environmental Sustainability: The Prospects for Market-Oriented Regulations

Mark Jaccard

Shifting toward a more environmentally sustainable path is one of the great challenges facing humanity. Although policy making in any domain is rarely easy, environmental sustainability appears to conflict directly with the material-growth tendency of economic systems. In the face of this formidable force, government sustainability initiatives often appear tentative at best. Indeed, for those who want society to take a more precautionary approach to environmental risks, development of government policy seems to move at a snail's pace or be so diluted when finally implemented as to have negligible effect.

For several decades, analysts have debated the merits of alternative policy mechanisms for achieving environmental targets. Increasingly, these debates refer to real-world evidence as governments experiment with different policy approaches, the relative merits of which are presented in studies that survey policy successes and failures around the world. One lesson from this growing literature is how difficult it is to design policies that satisfy the standard criteria of policy design – administrative feasibility, political feasibility, economic efficiency – while also being effective at inducing profound technological change that improves environmental quality or reduces environmental risks.

While much attention has recently focused on the potential for firms to voluntarily change their business practices for real environmental improvement, there are good reasons to be skeptical. Firms point to specific improvements in the efficiency with which they use material and energy inputs in a given process, or reductions in the rate at which they emit effluents and emissions. But such changes may simply reflect ongoing productivity improvements that, while impressive for an individual business activity, will not deflect the worsening trend of a particular environmental threat when all activity is measured in aggregate. Indeed, one risk is that with the current excitement about a possible change in business culture, governments will neglect their duty to ensure that all businesses operate under a set of rules and incentives that provide a consistent economic signal about society's goals for the environment. In this chapter, I focus on the development of a category of policies that provides an external stimulus to mobilize producers toward environmental sustainability.

The chapter is organized as follows. In the following section, I summarize the major challenges to environmental policy making. As an example, I focus on greenhouse gas (GHG) emission reduction policy, using Canada as a case study. Following that, I describe the well-known policy options – voluntarism, command-and-control regulation, economic instruments – and present some recent policy innovations, focusing especially on the policy approach that I emphasize in this chapter: market-oriented regulations. Although this approach is not well known, there are already some practical applications that allow me to make a preliminary assessment of its strengths and weaknesses. The next three sections are devoted to a detailed description of three prominent examples of this approach: the emission cap with tradable permits in the electricity sector, the renewable portfolio standard in the electricity sector, and the vehicle emission standard. In the conclusion, I reflect on the broader lessons for environmental policy making and comment on the general prospects for Canadian application of these types of policies, returning to GHG emission reduction as my example.

Policy Challenges for Sustainable Production: The Case of Greenhouse Gases in Canada

The goal of environmental sustainability presents special challenges for policy making. Three of these stand out.[1] First, it is difficult for people to connect their actions as consumers with the local environmental impacts that concern them as citizens. Many of these local impacts are hidden from view. For example, waste generation and disposal, such as landfills and industrial production, are generally located outside urban centres. Also, many byproducts of consumption, even the most potentially harmful, are unseen effluents and emissions with risks that relate to persistence and long-distance transport, effects that cannot be readily detected by experts, let alone the average citizen.

Second, the environmental effects of human activity are increasingly global and intergenerational. This extension makes it difficult for the current generation in one region to connect its pursuit of well-being and security with the potential repercussions for the well-being and security of future generations, and even for current inhabitants of the other side of the planet. The highly uncertain nature of these distant effects diminishes the public and political will to take action.

Third, because humans have evolved an economic system that sometimes treats the environment as a free and unlimited waste receptacle, there will be substantial transitional costs in shifting that system toward substantially lower flows of the waste by-products that pose environmental risk. These costs must be incurred at a time when the options for government have narrowed. Governments lack the financial resources to pay all of these costs from existing revenues. Regulations that raise costs are seen as heavy-handed. Tax reforms designed to change technologies and behaviours in ways that will reduce environmental risk, even reforms that do not increase the net tax burden, are met with antagonism by significant segments of the public and media in what is usually a polarized political climate. We have entered an era in which governments are reluctant to set rules or adjust costs, opting instead to suggest, lead by example, and form voluntary partnerships (see Pal 1997; Executive Resource Group 2001). Some wonder how this restricted power can provide the leadership and regulatory authority necessary to shift our economic system onto a more environmentally sustainable path.

The effort to reduce GHG emissions illustrates these three challenges. Many scientists believe that humans should stabilize and perhaps reduce atmospheric concentrations of GHGs over the next half-century in order to reduce the risk of climate change that might harm humans and ecosystems. Stabilization of concentrations in this timeframe means that humans would need to reduce total anthropogenic emissions by close to 70 percent over the next few decades. This represents a profound change in the current global energy system at a time when that system is expanding to meet the wants of comparatively wealthy people in industrialized countries and the needs of poor people in developing countries.[2]

How can governments mobilize the public and corporate effort that is needed? First, the connection between our actions as consumers and the resulting GHG emissions is not readily apparent to most people. The GHG emissions from the direct use of fuels – as in a vehicle or furnace – originate locally but they are not visible, and the indirect GHG emissions from our use of electricity usually originate unseen at a distant generating plant. Even more indirect and unseen are the GHG emissions resulting from the production and transport of the goods and services we consume. Second, much research suggests that the most significant climate change impacts from the GHG emissions of North Americans and Europeans will happen to people living in low-lying and ecologically fragile countries in other parts of the world, and to future generations. Third, the reduction in GHG emissions that scientists say is needed to stabilize atmospheric concentrations requires not only the rapid dissemination of existing technologies that can improve energy efficiency and provide low-emission energy supply, but also innovations that lead to dramatic technological change, and this could prove to be costly.

The issue of cost is especially important from a policy perspective. If the cost of substantial GHG reduction is low, industry and the public are more

Figure 8.1

Estimated GHG reduction cost curve for Canada: Kyoto target in 2010

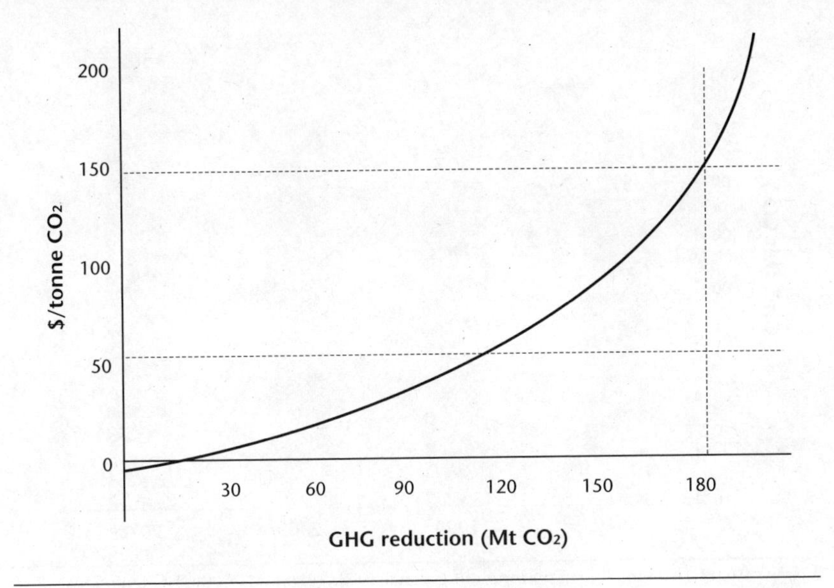

likely to support ambitious emission reduction targets. If the costs are high, the policy tension between short-term financial costs and hoped-for but uncertain environmental benefits over the long run will be significant.

Figure 8.1 provides an estimate made in 2000 of Canada's cost of reducing its GHG emissions by about 30 percent (180 megatonnes of carbon dioxide) by the year 2010, which is the target Canada committed to under the Kyoto Protocol in 1997. (The specific Kyoto target is 6 percent below Canada's 1990 emissions level, but this is equivalent to a 30 percent reduction from the business-as-usual emission projection for 2010.) The figure shows a marginal cost of $150 per tonne of carbon dioxide for reducing emissions by 180 megatonnes.[3]

Because technologies acquired in the years before 2010 would continue to operate beyond this period, net annual costs to 2022 are included in assessing the costs of GHG reduction induced by policies during the Kyoto period. When the net costs over the full twenty-two years are converted into present value (using a discount rate of 10 percent), the total cost of the 180-megatonne GHG reduction is $70 billion. This cost could cause a decline in economic output by the Canadian economy of about 3 percent by 2010, the equivalent of a one-year recession (Jaccard et al. 2002). More recent projections of current emission trends suggest that the required reduction could be as great as 240 megatonnes or even 300 megatonnes, implying

Figure 8.2

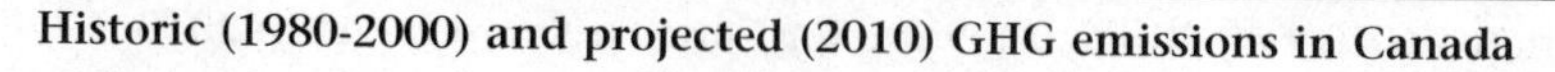

Historic (1980-2000) and projected (2010) GHG emissions in Canada

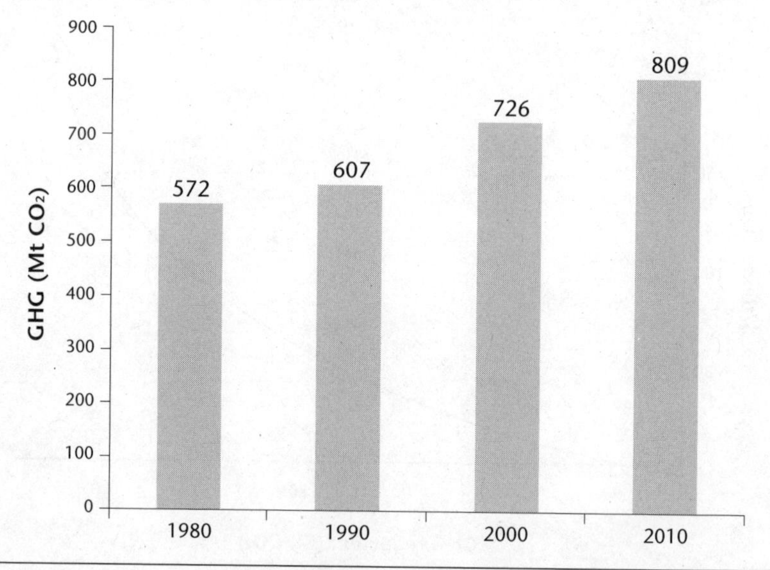

Sources: 1980 GHG emissions: Environment Canada 1999; 1990 and 2000 GHG emissions: Environment Canada 2002; 2010 GHG emissions: Government of Canada 2002.

an even higher cost of compliance with the protocol (Figure 8.2) (Government of Canada 2002).

Another way of interpreting these cost estimates is in terms of the energy prices resulting from the application of the economy-wide financial policy instrument – such as a GHG tax – necessary to stimulate the shift in technologies and behaviour by businesses and consumers that would reduce annual emissions by 180 megatonnes in ten years. The results imply electricity price increases ranging from almost zero in hydro provinces to 100 percent in provinces that generate most electricity from fossil fuels, and natural gas and gasoline price increases of 50-80 percent depending on the province (Jaccard et al. 2002).

If these high cost estimates are indicative, the objective of GHG reduction to reduce environmental risk provides a good example of the three environmental policy challenges listed above: the link between personal decisions and GHG emissions is not readily apparent; many impacts may be felt by people removed from us in space and time; and the costs of immediate action appear to be substantial. However, while the costs of a 30 percent reduction of GHG emissions by Canada appear to be high, these costs are a function of several factors that might, in turn, be influenced by policy. Important factors include the time frame for achieving environmental targets, the effect of policy on innovation, and the effect of policy on the long-term evolution of business and consumer preferences. Jaffe and colleagues (2002) explore these and other factors in their comprehensive survey of research into the relationship between environmental policy and technological change. They find considerable empirical support for the argument that the costs of environmental improvement are sensitive to the timing and choice of policy instrument. Some policies better match the pace of technological change to the natural rate of capital stock turnover, thus reducing compliance costs. Some policies are better at inducing research and development investment into those profound technological innovations that can substantially reduce the costs of environmental improvement in the long run. Some policies are better at influencing the preferences of businesses and consumers, and at achieving political acceptance. Some policies are better at mobilizing producers to transform the market as quickly as possible.

Policy Options for Sustainable Production: Conventional Approaches and Recent Innovations

In this section, I present standard criteria for evaluating environmental policies and then assess current and emerging policy options in terms of their performance against these criteria. My discussion focuses especially on policies that perform well in the short-term and that may mobilize producers to develop and commercialize environmentally desirable technology that will lead to reduced costs.

Policy analysts use several criteria to compare the performance of alternative policy approaches. I rely on four criteria: (1) effectiveness at environmental improvement, (2) administrative feasibility, (3) economical efficiency, and (4) political feasibility. All four criteria are considered in concert. A policy may be politically acceptable but ineffective in achieving the intended environmental target. A policy may be effective but not administratively feasible or economically efficient. While no policy performs perfectly against all four criteria, some may do better than others in this difficult balancing act.

To guide the comparison, I display policy options in Figure 8.3 on a continuum in terms of their *degree of compulsion*. I use this term to express the extent to which behaviour is required by an external force. A fully compulsory policy specifies exactly what must be done, and non-compliance incurs severe penalties. This type of policy offers no flexibility for those to whom it applies. A policy is fully non-compulsory if the business or consumer can opt to do nothing without facing any negative consequences. Policies near the middle of the continuum would have some degree of compulsion while also allowing some flexibility to businesses or consumers in their choice of how to respond to the policy.[4]

Figure 8.3

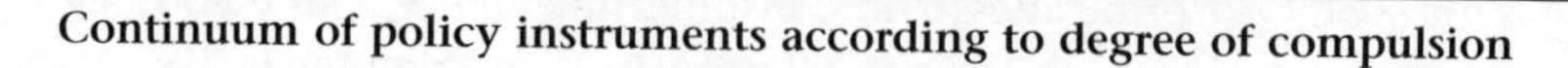

Continuum of policy instruments according to degree of compulsion

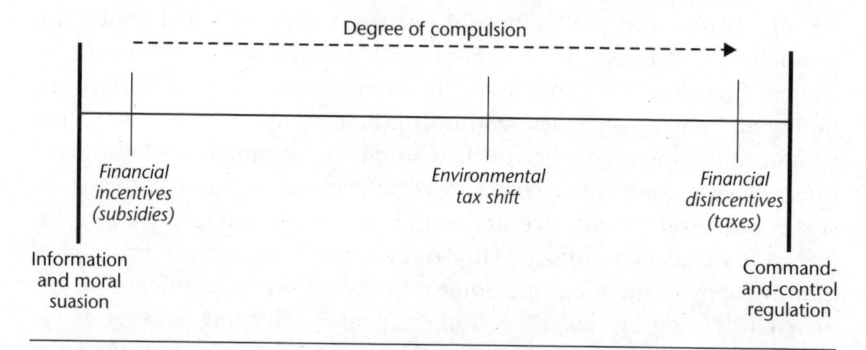

The *voluntary approach* – labelled "information and moral suasion" – is situated on the left side of Figure 8.3. This includes policies in which government facilitates or initiates the development and dissemination of information (research and development, advertising, labelling, certifying, providing demonstration projects) that might influence the decisions of households, firms, and perhaps other levels of government. Government would provide information that educates businesses or consumers about financial gain from actions that improve the environment, or that convinces them to undertake some costly actions for moral reasons, such as the desire to contribute to a cleaner, environmentally sustainable world. Finally, government would lead by example, taking actions where it has direct control (public lands, buildings, vehicles, employees, publicly funded research and development, publicly owned corporations) and hoping that others will follow or form partnerships with it for moral and financial reasons (Dietz and Stern 2002).

From a political feasibility perspective, the left side of the policy continuum scores high in that there is no government coercion. Voluntary policies are sometimes criticized for ineffectiveness, however, and this can foster political opposition from those who want stronger actions to protect the environment. There has been little research into the effectiveness of voluntary approaches to environmental improvement (Harrison 1999). In a recent survey of the literature on voluntary environmental policies, Khanna (2001) was able to identify only a couple of empirical studies examining policy effectiveness, neither of which showed that such programs had much effect. Similarly, the Organisation for Economic Co-operation and Development (OECD) recently concluded that the "environmental effectiveness of voluntary approaches is still questionable." It added that "the economic efficiency of voluntary approaches is generally low" (OECD 2003, 14). In the specific case of GHG emission reduction, Bramley (2002) suggested that the Voluntary Challenge and Registry (VCR) of the Canadian government is largely ineffective in achieving GHG reduction in industry, a conclusion based on the evolution of aggregate industrial emissions (24 percent increase in the period 1990-2000), lack of program coverage (less than 55 percent of total industrial emissions), and case studies of the target-setting and emission-accounting practices of some individual firms. The VCR was just one of several programs launched by the Canadian government in 1993 as part of its National Action Program on Climate Change, the objective of which was to reduce national GHG emissions to their 1990 levels by 2000. Figure 8.2 shows what happened instead, as emissions climbed even faster than during the previous decade to a level 20 percent higher than 1990 levels by 2000.

Financial incentives – such as grants, low-interest loans, tax credits, and other forms of subsidy – can be effective in reducing the ultimate costs of environmental improvement, especially when directed toward research and development, and new product commercialization (Carraro and Soubeyran 1996). Because they are not compulsory for recipients, I situate financial incentives on the left side of the continuum in Figure 8.3. They are, however, somewhat to the right of purely voluntary programs because government must acquire most of the necessary funds in a compulsory manner (income taxes, corporate taxes, reducing other expenditures). Thus, Canada's federal and provincial governments struggle to find funds for GHG reduction in the face of competing claims for tax reduction, greater health care spending, and an increased military budget (at the federal level). While the Canadian government has and will continue to provide financial support for GHG reduction, its contribution cannot cover more than a small percentage if the costs of its GHG reduction target are as high as the evidence presented here suggests.

Traditional *command-and-control regulations,* on the right side of Figure 8.3, mandate specific requirements for emissions, technologies, operating practices, and buildings and other structures. Non-compliance incurs severe financial penalties and legal sanctions. This approach dominated environmental policy in the 1970s and is still predominant today. Businesses and economists criticize the regulatory approach as economically inefficient, however, if participants are required to behave identically (same technology choice, same per unit emission level), even though some may be able to do more with lower expenditures than others. Also, prescriptive regulations do not promote innovation that might reduce long-term costs (Magat 1978). Businesses have therefore been fairly successful in convincing the public, politicians, and even environmentalists that a strict regulatory approach is an unnecessary drag on the economy, and this explains in part the recent interest in voluntary and other approaches. With respect to GHG emission policy, the Canadian government has not promoted command-and-control regulations as a policy to drive profound technological change, although it has used energy efficiency regulations to eliminate some of the most polluting technologies and practices (vehicles, appliances, industrial equipment, and buildings).

Financial disincentives, such as environmental taxes, are unit charges for emissions. While situated on the right side of the continuum in Figure 8.3, financial disincentives are not as compulsory as traditional regulation because they do not specify a particular action; the business or consumer chooses between no change – and paying whatever environmental charges are due – and investing in emission reduction in order to pay lower charges. As originally conceived, the unit emission charge should reflect the monetary value of the incremental damages caused by the emission, but this may not be achieved in practice as governments are constrained by considerations of political feasibility when setting the charge level.[5] Indeed, to achieve political feasibility, even the lowest level of environmental taxes must overcome the public's association of taxes with coercive government, and with the frequent assumption that taxes are motivated only by government's revenue needs. Federal and provincial governments in Canada have frequently expressed their unwillingness to consider financial disincentives as a policy option for GHG reduction.

As the foregoing discussion suggests, progression from left to right along the continuum in Figure 8.3 is associated not only with increasing compulsion but also with decreasing political feasibility. Firms and households want to retain as much freedom of action as possible. Politicians are well aware of the potentially strong reaction to policies on the right side of the continuum – or they soon learn from painful experience. Unfortunately, the continuum appears to correlate with another dimension: policy effectiveness. Non-compulsory policies lack evidence of effectiveness, whereas command-and-control policies – even if they can be economically inefficient – have proven to be effective in many cases in achieving environmental objectives. The challenge, therefore, is to develop policies that offer the best compromise in terms of these conflicting criteria.

One recent approach is to negotiate a *voluntary environmental agreement* (in the form of a contract, memorandum of understanding, or letter of intent) between an industry and government that specifies the industry's commitment to meeting a particular environmental outcome, such as a maximum industry-wide level of emissions. Depending on one's perspective, this approach can be seen as non-compulsory or compulsory. Industry and government may prefer the appearance of voluntary action even where both recognize that non-compliance might eventually require stronger policies such as regulations or financial disincentives. Thus far, however, there is no research demonstrating that voluntary environmental agreements can be distinguished from the general voluntary approach shown in Figure 8.3. Karamanos (2001) identified sixty-six voluntary environmental agreements but noted a lack of research assessing the effectiveness of this approach. Besides the concern with effectiveness, economic efficiency may be jeopardized if less competitive sectors (monopoly electrical utilities, for example) sign and implement agreements (being able to pass on the costs to captive customers) while more competitive sectors do not, even though the incremental costs of environmental improvement may be lower for the latter.

Another suggestion is to combine financial incentives and disincentives in what is referred to as *environmental tax shift* or *ecological fiscal reform.* This approach is revenue-neutral in that all revenues from environmental taxes are recycled as reductions in other taxes (preferably ones that are unpopular or economically inefficient) or as a rebate that does not distort the signal provided by the tax (Durning and Bauman 1998; NRTEE 2002). Modest environmental tax shifts include deposit/refund schemes and offsetting tax adjustments on low- and high-efficiency vehicles (called a feebate), while more ambitious initiatives under consideration, especially in Europe, involve the application of GHG tax revenue to reduce government payroll charges, income taxes, or other broad levies (Svendsen 1998; Svendsen et al. 2001). Given that many economists and environmentalists support environmental tax shifting, it may yet play a dominant role in GHG policy. It must overcome, however, the considerable suspicion among the public and media about government claims of revenue-neutrality. Politicians are reluctant to cause large price increases in essential commodities like energy, even if other taxes decrease as a consequence. Because of this problem with political feasibility, environmental tax shifting might initially play a consolidating rather than leading role, meaning that only modest tax changes would be experimented with in support of other policies that play the dominant role in driving long-term technological change.

Another fairly recent policy innovation is called *emission cap and tradable permits* (ECTP). Government sets a maximum level of emission (a cap), then allocates tradable emission permits to all emitters covered by the program. Usually the permits decrease in number or value over time, gradually lowering the aggregate emission cap. The ECTP is a form of regulation in that the aggregate emission cap cannot be exceeded, participation is compulsory, and penalties are severe for emissions in excess of permits. The policy is unlike traditional command-and-control regulation, however, in allowing each participant to choose by how much they will reduce emissions, if at all, and therefore whether they will buy or sell in the emission permit market. Because of these dual characteristics – regulatory, yet market-based – I refer to this policy approach as one form of *market-oriented regulation* (also known as a *quantity-based market instrument*).[6]

Government could apply an economy-wide ECTP policy for GHG abatement, allocating permits to final consumers or upstream producers of GHG-emitting products. This economy-wide application is similar to a GHG tax because the policy motivates emission reductions by increasing the price of GHG-intensive energy commodities and, as a consequence, the price of final goods and services that depend on these inputs. The final price of energy commodities will reflect their conventional cost of production plus the market price of tradable emission permits. To achieve an equivalent level of emission reductions, the market price of GHG permits should end up at about the same level as the required GHG tax rate (Jaffe et al. 2002). If government allocated the GHG permits by annual auction, the ECTP would generate government revenues comparable with the revenue from a GHG tax. For an outcome similar to environmental tax shift, government could use the auction revenue to reduce other taxes. If, instead, government allocates permits based on emissions in a previous period (grandfathering), the distributional effect of the ECTP would be similar to a GHG tax strategy in which tax revenues are recycled back to participants in proportion to their GHG tax payments in the initial year of the program.

Being similar to a GHG tax, the economy-wide ECTP shares some of its strengths and weaknesses as a policy tool. Like GHG taxes, the economy-wide ECTP should be effective and is economically efficient. Similarly, the ECTP would result in dramatic energy price increases if the cost of GHG reduction is as high as suggested by the research reported here. Applied as the key driver of long-term GHG reduction, the ECTP will therefore face the same political feasibility challenge as GHG taxes. Unlike taxes, the ECTP may be more administratively difficult if applied as an economy-wide policy, as this would require the creation and operation of an administratively feasible permit-trading mechanism.

While the concept of market-oriented regulation has generally been applied to emissions, in this chapter I explore the potential for applying the principles behind this approach more broadly, to include its application to technologies, forms of energy, product characteristics, or any attribute of technological change that can be linked to long-term environmental improvement. I also consider how to design this approach in ways that improve its political feasibility and minimize its potential negative impacts on economic efficiency. In the next three sections, I describe three cases in which sector-specific, market-oriented regulations have been applied, and then in the final section I draw lessons from these and discuss the potential for extending the approach.

Case 1: Sulphur Dioxide Emission Cap with Tradable Permits

ECTP programs have been implemented by the US Environmental Protection Agency since the mid-1970s. The most ambitious and noteworthy application was established under amendments to the US Clean Air Act in 1990.[7] This amendment set 1995 as the start date for a program that allocated tradable permits for sulphur dioxide emissions to 110 electricity generation plants initially, expanding the program to several hundred plants in the second phase beginning in 2000. The tradable permits (emission allowances) were allocated on the basis of historical emissions, but additional permits were auctioned in order to ensure access for new market entrants. Each year the total number of permits decreases. For emissions in excess of its permits, a plant would be subject to a fine of $2,000 per tonne of sulphur dioxide.

The key element of the program is the provision for emission permit trading, which is facilitated by several components. An offset policy ensures that the trading program is sensitive to the needs of special areas that already exceed local air pollution standards. Thus, new plants must offset by an additional 20 percent any new emissions that they would introduce to non-attainment areas. A bubble policy allows trading between all emission sources that are within a particular area as long as the total bubble objective of the regulator is met. A banking policy allows emitters to save emission allowances for use in future years. Finally, the policy was implemented gradually, with considerable warning, in order to provide the opportunity for electricity generators to implement emission reduction measures as cost-effectively as possible.

While it is still too early to assess comprehensively the sulphur dioxide cap and trade policy, the costs of emission reduction have been much lower than originally expected. Early estimates suggested that permits might have a trading value of $1,000 per tonne of sulphur dioxide, resulting in an annual cost of about $10 billion to meet the emission reduction target. In the first phase, from 1995 to 2000, the price ended up being about 10 percent of this, with a total annual cost of less than $1 billion. Some of this cost reduction is directly attributable to permit trading between companies, while some is a result of both a fall in the cost of transporting low-sulphur coal due to railroad deregulation and a decrease in the cost of desulphurization equipment as the lengthy warning about the policy start-date gave equipment manufacturers time to innovate.

Given this apparent success, the ECTP approach might appear to be ideal for many types of environmental issues. The sulphur emission problem had certain characteristics that were favourable to this particular policy approach, however:

- The relatively small number of homogeneous firms (utilities) in the electricity sector ensured the administrative feasibility of designing and implementing the sulphur program.
- The regional nature of the acid rain problem ensured that there were sufficient firms in the affected area so that the program might yield

economic efficiency benefits from its permit trading arrangements. The regional nature of the issue, however, also meant that international co-ordination and trading (in contrast with GHGs) were not required as part of the policy, again contributing to administrative feasibility.

- In most jurisdictions, electricity sector prices are regulated by utilities commissions to reflect average production cost, meaning that the high incremental cost of emission reduction for some plants would be averaged with the costs of all other relevant plants in setting rates. Political acceptability is easier to achieve when dramatic rate increases can be avoided, even though this form of pricing is likely to be inconsistent with the economic efficiency objective (consumers do not see the high incremental cost of electricity production when making consumption decisions).

For some time, governments have considered applying ECTP policy to the goal of reducing GHG emissions.[8] A major challenge, however, is the possibility of unbearably high GHG permit prices leading to politically unacceptable energy price increases. The use of ECTP as the sole mechanism for achieving Canada's GHG abatement commitment under the Kyoto Protocol would lead to the same kind of price shocks as the $150 GHG tax illustrated in Figure 8.1. To address this concern about cost uncertainty, policy makers can set a permit price ceiling by offering an unlimited number of permits at this price. In effect, this converts the ECTP into a hybrid price/quantity policy that functions like a quantity regulation when the permit price ceiling is above the permit trading price, and like a tax when it is not. This approach to the ECTP has attracted considerable interest in recent policy discussions in Europe, the US, and elsewhere (Jacoby and Ellerman 2004; Pizer 1999). The permit ceiling price is described as a "safety valve" in that it protects industry from the uncertainty that government will select a cap for which compliance will cause major economic repercussions. In this regard, an additional innovation is to start with a low permit price ceiling and issue a schedule indicating its gradual climb, so that industry has time to adjust but will be aware of future GHG cost liabilities when deciding about major new investments in the present.

The ECTP specifies a particular market outcome with respect to a maximum amount of emissions in future time periods. This type of market-oriented regulation can be modified, however, to specify a minimum market share for particular technologies or forms of energy. The rationale is that because some less-polluting devices and forms of energy are associated with new, unconventional technologies, these need a guaranteed minimum market share (or guaranteed niche market) in order to achieve the economies of scale and economies of learning that substantially lower costs over time. I refer to these as *niche market regulations* – a category of market-oriented regulation, like the ECTP. The second and third cases below apply this principle to the choice of energy form (electricity) and power platform (vehicles), respectively.

Case 2: Renewable Portfolio Standard in Electricity Generation

With the exception of hydro and nuclear power in a few countries, fossil fuels, especially coal, dominate worldwide electricity production. Their combustion is linked with local smog, regional sulphur emissions, and global greenhouse gases. Renewable sources – sunlight, wind, organic matter, running water, wave action, tidal flows, and geothermal energy – can usually generate electricity with much less environmental impact, but fossil fuels can produce electricity at a lower financial cost than most renewable technologies if the externality damages from fossil fuels are ignored.

As part of environmental sustainability strategies, governments favour renewable electricity generation technologies. One intended consequence is that because renewable sources of electricity are mostly associated with new technologies, their costs should fall substantially as market diffusion leads to learning and economies of scale in manufacture. Research on technology innovation and diffusion indicates that technology costs undergo a dramatic decline once the scale of production surpasses critical thresholds (Grubler et al. 1999). Policy support for renewables may always be required, however, if the prices of more polluting alternatives are not compelled to fully reflect the full environmental costs they cause – and this might not occur because of the political acceptability constraints on economic instruments. The types of policies that have been developed to support renewables include:

- subsidies for research and development
- capital subsidies to renewable technology investments (grants, low-interest loans, tax credits)
- electricity purchase price subsidies
- information and voluntary programs (demonstration projects, advertising and training, renewable resource assessments, encouragement of voluntary agreements by producers, and green tariffs that give consumers the option to pay a premium to ensure renewable electricity production to meet their needs)
- reduction or elimination of subsidies to non-renewable sources of electricity
- additional restrictions and charges to non-renewable electricity production.

These policies have had some success. Perhaps most significant have been electricity purchase price guarantees for renewable electricity, which contributed to the dramatic decline in the cost of wind-generated electricity to the point where today wind is competitive with fossil fuels in some locations.

During the 1990s, the *renewable portfolio standard* (RPS) emerged as a new policy instrument for supporting renewable electricity generation. This niche market regulation requires that the sellers of electricity procure a minimum percentage of the electricity they sell from renewable sources.[9] It does not specify the production choice of each producer, only the aggregate market outcome. Each producer decides whether it will produce some electricity itself from renewable sources or somehow combine with, or purchase from, a renewable electricity provider. Thus, if markets function effectively, producers will trade among themselves – as with the tradable emission permits – and only those with the lowest costs will generate renewable electricity for the entire market.

Four characteristics help explain the emerging interest in the RPS:

- The RPS maintains a continuous incentive for renewable producers to seek cost reductions (economic efficiency) and can be designed to ensure that these cost reductions are passed on to consumers (equity). This is achieved by mechanisms that establish continuous cost competition among renewable producers for their share of the RPS; there is no guaranteed electricity price, only a guaranteed market share for renewables.
- Because each purchaser of electricity is paying a blended price, consisting of the higher-cost renewables along with the dominant, lower-cost conventional supply, the RPS has a negligible effect on current rates. RPS targets are modest initially, giving time for the market to adjust and for competitive pressures and diffusion to drive down the cost of renewables.
- Because the RPS ensures the attainment of a specific market share, it can be directly linked to government policy objectives. Governments can set the RPS requirement as part of a package of policies to meet an environmental target such as GHG abatement.
- The RPS minimizes government involvement relative to other measures. The government's budget is not implicated because customers pay producers directly the extra financial cost of renewables, and the selection of winning bids can be left to market forces (competitive bidding) instead of government evaluation.

The RPS has been adopted by Australia, almost twenty US states, and several countries in Europe. It has also been under consideration in legislation before the European Union and the US Congress. Because the RPS has only recently emerged as a policy tool, there is insufficient experience to provide an empirical evaluation of its performance relative to key alternatives, such as subsidized feed-in tariffs. Such an evaluation is complicated by the fact that jurisdictions that have adopted the RPS have retained at least some of their other policies supporting renewables.

Critics have argued that the RPS is flawed because it forces the adoption of particular forms of energy, even though these are but indirect means to an end, such as environmental improvement. In so doing, the RPS may delay or prevent the development and diffusion of technology innovations that would more economically achieve the same objective. If, for example, a key motive for the RPS is to abate GHG emissions, a policy focusing directly on these emissions – such as a GHG tax or emission cap and tradable permit system – would open the door to technologies such as the production of hydrogen from coal with capture and permanent storage of all potentially harmful by-products in geological formations. Conceivably, this or other alternatives to renewables could meet all of the public objectives behind the creation of the RPS at a much lower cost, freeing resources for other social objectives.

Advocates of the RPS have countered that non-renewable resources such as clean coal cannot possibly outperform renewables in providing affordable energy while also satisfying long-term environmental, security, and social objectives. A less exclusionary argument in favour of the RPS is that renewables are likely to be part of a future energy mix that might one day also include cleaner fossil fuel and nuclear technologies, but that renewables need special help at this stage to pass critical diffusion thresholds so that their costs will fall. If the RPS includes tradable certificates and other flexibility mechanisms, it reduces the policy cost and again increases the likelihood of political acceptability.

The next case shows how niche market regulation has been applied to technology choice in the personal vehicle sector of the economy.

Case 3: Vehicle Emission Standard

Personal vehicles are responsible for a considerable share of urban air pollution and are also significant contributors to GHG emissions. In industrialized countries, Los Angeles is renowned for its air quality problems and California has been a leading jurisdiction in vehicle-focused policies for environmental improvement.

There are three general actions for reducing vehicle-based urban air pollution: (1) decrease the need for vehicle use by fostering alternatives such as reduced travel demand and increased walking, cycling, and use of public transit; (2) increase the market share of high-occupancy vehicles; and (3) produce cleaner vehicles either through greater efficiency, better pollution control equipment, or switching to cleaner fuels and technologies.

Governments have fostered all of these actions through various policies, including information campaigns to influence choice of vehicle and mode of mobility, efficiency and emission regulations, taxes, and subsidies to public transit. Efficiency and emission regulations are especially common throughout the world (Faiz et al. 1996). In the US, most states conform to the fleet average emission regulations of the Environmental Protection Agency, although California has had special authority since the 1970s to set more stringent standards because of its exceptional air quality concerns.

Until the 1990s, the prospects were poor for a dramatic technological change toward extremely low or even zero-emission vehicles, with manufacturers claiming that these would be too costly to build and would not satisfy consumer demands for acceleration, horsepower, safety, and range. In 1990, however, the California Air Resources Board radically changed technology expectations and policy design with the adoption of a *vehicle emission standard* (VES) that regulated a niche market for zero- and low-emission vehicles.[10] California's VES required that such vehicles achieve minimum market shares of new vehicle sales by specified dates over the next two decades. Individual manufacturers of vehicles are expected to meet the standard as a fleet average of retail sales in California, meaning that they could exchange credits among themselves so that the total California fleet met the VES even if individual manufacturers fell short.

The California VES seems to have played a pivotal role in the development of new options for vehicle platforms, including gasoline/electric hybrids, battery/electrics, and fuel cell. Indeed, manufacturers appear intent on outcompeting each other to capture this new market, as reflected by recent research funding, commercialization efforts, and marketing strategies. While the state's VES was modified in 1996 and again in 1998, its essential character and performance requirement has not changed since its inception. Moreover, the California legislation has been adopted by twelve other states, with provisions in New York, Massachusetts, Vermont, and Maine to automatically adjust their standards to any changes made in California (these states and California together account for 20 percent of the US vehicle market). The policy has also had a significant effect on technology developments by vehicle manufacturers in Europe and Japan, although no other country has copied it yet.

The VES has not been implemented without difficulties, especially in the debates, negotiations, and legal challenges surrounding the zero-emission vehicle standard (Pilkington 1998; Kemp 2004). The initial zero-emission requirements were for 2 percent market share in 1998 and 10 percent in 2003 (measured as a percentage of 1992 new vehicle sales). The California Air Resources Board, however, reduced these requirements and added flexibility provisions in a negotiated agreement with automobile manufacturers in 1996 because of setbacks in the development and commercialization of battery/electric vehicles, which were initially assumed to be the favoured zero-emission option. With emerging evidence that fuel cell vehicles might triumph over battery/electric in meeting the zero-emission requirement, albeit in the more distant future, and the realization that gasoline/electric hybrid vehicles might soon achieve widespread diffusion, the board conceded that extra sales of hybrids could compensate for missing the zero-emission targets. In 2004, California passed legislation to add GHG requirements to the local air pollution focus of the original policy. Other countries, such as Canada, are looking at the possibility of adopting the California VES with these new GHG provisions.

The VES is similar to the RPS in that the environmental objective is the indirect outcome of a flexible, technology-focused regulation applied to producers. In this case, the standard is applied to the characteristics of a durable good (automobiles), instead of the production process of a commodity (electricity).

Several factors explain the attractiveness of the VES:

- The policy is sector-specific, improving the prospects for administrative feasibility.
- By allowing trading among manufacturers, and a long time frame for achieving targets, it reduces the cost of technological change.
- The policy does not require dramatic increases in the price of vehicles or fuel to stimulate the private R&D and commercialization effort needed for dramatic technological change in vehicles, thus improving the prospects for political acceptability; small increases in the costs of conventional vehicles should be sufficient to cover the extra R&D investment and to subsidize the price of new, zero- and low-emission vehicles.

California's VES has been described as a huge failure and a dramatic success. Critics claim that it failed to produce a zero-emission vehicle in the intended time frame and wasted public and private money in the process. Supporters claim that it pushed automobile manufacturers worldwide to design and commercialize a technological transformation that is providing real benefits in terms of commercially attractive low-emission vehicles. They argue that California's regulators demonstrated foresight in first pushing for different categories of low- and zero-emission vehicles, and then adaptability in recognizing the potential for hybrid/electric vehicles to meet the short-term emission goal while allowing time to ascertain whether battery/electric or fuel cell vehicles will prevail in fulfilling the long-term zero-emission goal.

Policy Applications: Canadian Prospects

The three types of market-oriented regulations described in this chapter are different in various respects, yet similar in terms of how they address the significant constraints on environmental policy making. They reflect a new trend in environmental policy making that focuses on the strong incentives needed if we are to bridge the gap between our understanding of the technological change that is required and achievable, on the one hand, and our ability to mobilize producers to make that change, on the other. The following common characteristics of all three policies highlight their potential to perform well against the four policy evaluation criteria of effectiveness at achieving environmental targets, administrative feasibility, economic efficiency, and political feasibility.

- As aggregate regulations with substantial penalties for non-compliance, all three policies should be effective in achieving directly or indirectly the environmental target. This is in contrast to the performance uncertainties associated with voluntary and subsidy policies. The ECTP is directly linked to an environmental target. Because the RPS and VES regulate technologies, their environmental effects are less direct and therefore less certain; growth in electricity sales or vehicle sales could offset in part the aggregate environmental benefits of the policies. However, countries like the Netherlands and Britain tie the design of their RPS policies closely to their national targets for GHG reduction. In the Netherlands, the certificates that enable renewable electricity trading between electricity producers are also linked to that country's carbon tax policies, improving the prospects of cost consistency throughout the economy. Likewise, California links its slate of vehicle emission requirements directly to the attainment of its local air quality objectives and more recently its GHG reduction goals.
- The three policies are designed to provide long-term signals to producers that will motivate the development and dissemination of new, cleaner technologies without significantly affecting the prices currently paid by consumers. With the ECTP policy for sulphur emissions, producers faced the marginal costs of reducing emissions from electricity production, but this was in an industry whose retail rates are regulated to equal its average cost of production. Likewise, the price to electricity consumers in jurisdictions with RPS is a weighted average of the conventional electricity production costs with the incremental renewable electricity production costs, with a negligible effect on prices. The VES is the same in that producers have an incentive to subsidize if necessary the higher cost of lower-emission vehicles from the sales of conventional vehicles. The minimization of price impacts is an important component in achieving political feasibility, especially during the critical stage of developing low-emission technologies that would eventually offer an alternative to businesses and consumers.
- The three policies focus on the link between new product commercialization and the mass dissemination necessary to realize substantial cost reductions from economies of scale and economies of learning. The sulphur emission cap, the RPS, and the VES are designed to ensure more rapid

attainment of a minimum market share for targeted new technologies so that the costs of these technologies will fall more quickly.

- The three policies sustain a continuous competitive pressure for producers to find lower-cost ways of meeting the environmental or technology target. In this sense, the policies operate like any economic instrument, and the result should be ongoing cost reductions and reduced impacts on consumer prices. Again, this improves the prospects for public and corporate acceptance of the environmental target and the implementation policy.
- All three policies minimize the direct intervention of government in detailed decisions of the market. Given the all-too-frequent suspicions that government intervention will be heavy-handed and compulsory, this too may improve the prospects for political feasibility.
- To the extent that the policies affect the products offered by producers, they may also influence their marketing strategy, encouraging a shift toward green marketing. This may have profound implications, given that sellers are more effective than government at advertising. The VES now motivates automobile manufacturers to market cleaner cars, and this change in perspective may ultimately influence consumer preferences. This potential is less with the RPS because it is usually directed at electricity wholesalers rather than retailers. There are cases, however, where retail electricity competition in concert with RPS will enable producers to cover some of their costs by inducing customers to pay a premium for cleaner power. It may even be possible under retail competition for producers to market their electricity in terms of the amount of sulphur or GHG emissions per kilowatt-hour. Also, once producers have made a commitment to manufacturing something in large quantities (sulphur scrubbers, renewable electricity, low-emission vehicles), it is eventually in their interests to continue the market transformation so that older technologies and their production chains are phased out of the market.
- By focusing on one sector of the economy, all three policies minimize the complexities of policy design and thus improve the prospects for administrative and political feasibility. There are fewer parties to consider, meaning lower transaction costs in policy design and implementation. The sector-specific approach has a greater risk of economic inefficiency, however, because technological change and market transformation in some sectors may be very different from those in others in terms of the marginal cost per unit of environmental improvement. This is why economists generally prefer the economy-wide application of economic instruments like taxes and tradable permits. If policy makers nonetheless pursue sector-specific policies because of their other advantages, then research and ongoing monitoring is required to ensure the selection of technology, emission, and energy targets for which the incremental costs of

environmental improvement do not diverge dramatically from one sector to the other. Also, once sector-specific, market-oriented regulations have initiated technological change and market transformation, it becomes easier for governments to apply GHG taxes or an economy-wide ECTP, because then consumers have an option to purchase renewable electricity or low-emission vehicles whose cost has fallen closer to that of the conventional alternatives.

Because the market-oriented regulations described here are relatively new, it is still early to speculate on the full potential of this approach to environmental improvement. It is easy, however, to conceive of the application of niche market regulations to various sectors of the economy. If society wants more environmentally friendly buildings, government could require a minimum market share for the most advanced structures and allow builders to trade among themselves in meeting that target. If society wants to understand the prospects for zero-emission uses of fossil fuels, government could require a small, minimum market share for carbon capture and storage, and allow fossil fuel producers (coal, oil sands, conventional oil, natural gas) to trade among themselves or form a single conglomerate in meeting the target.

A flourishing of niche market regulations may be attractive for those who are convinced that the technological change we need for environmental sustainability is profound, and that producers need targeted stimuli in almost every sector of the economy. But each niche market regulation in a different sector increases the risk that society is taking the higher-cost path toward achieving its environmental goals. In particular, economists will argue that an economy-wide ECTP is preferable to technology- and energy-specific niche market regulations because it provides a consistent price signal to all sectors of the economy, whereas the niche market regulations could drive high-cost actions in one sector while low-cost actions in another sector are neglected.

A compromise approach would apply an economy-wide ECTP in concert with only a few niche market regulations, these latter focusing on key technological developments where individual businesses are unlikely to explore promising technological paths because of the financial risks. A recent policy-modelling study on long-term GHG abatement for Canada took this approach by matching a widely applied ECTP with three niche market regulations: RPS, VES, and a carbon capture and storage requirement (Jaccard et al. 2004). The cap of the ECTP was set to gradually fall, while its permit price ceiling would gradually rise so that at some future date, the ECTP would supplant the need for the niche market regulations. The purpose of the latter was to make sure that new technologies had passed through the critical commercialization and early diffusion stage in a timely fashion so that they would be ready to compete in the world of rising GHG emission charges (in the form of a rising permit price ceiling). According to this study, the impacts on the Canadian economy would be small (less than 1 percent of GDP), even for substantial GHG abatement (250 megatonnes of annual carbon dioxide emissions) over a thirty-five-year time period.

Canadian governments have been slow to adopt market-oriented regulations to mobilize producers toward environmental sustainability, but this may be changing. As part of its more recent GHG abatement efforts, the federal government has proposed an ECTP for large final emitters, which covers all energy-intensive industries, including the energy sector (electricity, refining, oil and gas production, and coal). If implemented, this proposal is likely to have a permit price ceiling.

The federal government in 2005 was closely studying the California VES, although it ultimately opted for a voluntary agreement with automobile manufacturers. The federal government has less jurisdictional authority to implement an RPS, but it is encouraging the provinces to take the initiative, and some have done so. British Columbia has a 50 percent RPS for new electricity generation. This is a voluntary RPS, but it is being applied by BC Hydro, a Crown corporation. While these tentative efforts are encouraging, the potential for this new policy approach to balance the diverse criteria of policy design suggests that it merits a much more serious examination in Canada by governments and interest groups. The great challenges of environmental policy making require that much at least.

Notes

1 Numerous works describe the special challenges of environmental policy making. The following are a selection from the much larger list: Loehman and Kilgour 1998; Lutz 1999; Prugh 1995; Sagoff 1988; Sandler 1997. For a Canadian perspective, see Boardman 1992, and Hessing and Howlett 1997.

2 Anthropogenic GHG emissions are not just energy-related, but I focus on these in this discussion. The general policy implications can be equally applied to reduction strategies in agriculture, forestry, and other GHG-causing activities.

3 This estimate was generated using an energy-economy model that is a hybrid of the economics-focused top-down models and the engineering-focused bottom-up models (Jaccard et al. 2002). It tends to yield cost estimates between the higher estimates of the former and the lower estimates of the latter.

4 Some policy analysts prefer other criteria, or would characterize these four differently. For example, equity could replace political feasibility. In terms of GHG policy, however, equity is difficult to define. Are equal per-capita policy costs equitable? Are equal regional costs equitable? Are costs that reflect each person's contribution to GHG emissions equitable? Political feasibility here means that politicians can find sufficient support to implement a policy. Thus, one can envision situations where a policy, such as GHG taxes, might pass the first three criteria while failing the test of political feasibility, or where a policy, such as voluntary action, might fail two or three criteria but pass the test of political feasibility.

5 When presenting the concept in the 1920s, Pigou (hence the term "Pigouvian taxes") noted that the tax should be equal to the marginal value of the damages in order to provide

economically efficient price signals. For an overview of economic instruments, see Kosobud and Zimmerman 1997 and Andersen and Sprenger 2000.

6 Some economists place ECTP under the broad category of economic instruments alongside financial incentives and financial disincentives. I prefer the separate category of market-oriented regulation because of the strong regulatory character, albeit at an aggregate level, of the emissions cap.

7 For a more detailed description, see Tietenberg 1998 and Stavins 1998.

8 In 2005 Europe instituted a cap and trade system for controlling greenhouse gases that applies to large industrial emitters, including electric utilities.

9 For a more detailed description, see Jaccard 2004; for a survey, see Berry and Jaccard 2001. The policy could be equally applied to the buyers of electricity instead of the sellers, as it has at one point in Denmark.

10 The California Air Resources Board is one of six quasi-independent regulatory boards operating under the authority of the California Environmental Protection Agency. Details of its policies are found on its website (http://www.arb.ca.gov).

References

Andersen, M., and R.-U. Sprenger, eds. 2000. *Market-based instruments for environmental management.* Cheltenham, UK: Edward Elgar.

Berry, T., and M. Jaccard. 2001. The renewable portfolio standard: Design considerations and an implementation survey. *Energy Policy* 29 (4): 263-77.

Boardman, R., ed. 1992. *Canadian environmental policy: Ecosystems, politics and process.* Toronto: Oxford University Press.

Bramley, M. 2002. *The case for Kyoto: The failure of voluntary corporate action.* Drayton Valley, AB: Pembina Institute.

Carraro, C., and A. Soubeyran. 1996. Environmental policy and the choice of production technology. In *Environmental policy and market structure,* ed. C. Carraro, Y. Katsoulacos, and A. Xepapadeas, 151-80. Dordrecht, Netherlands: Kluwer Academic Publishers.

Dietz, T., and P. Stern. 2002. *New tools for environmental protection: Education, information and voluntary measures.* Washington, DC: National Academy Press.

Durning, A., and Y. Bauman. 1998. *Tax shift.* Seattle: Northwest Environmental Watch.

Environment Canada. 1999. *Canada's greenhouse gas inventory: 1997 emissions and removals with trends.* Ottawa: Environment Canada.

–. 2002. *Canada's greenhouse gas inventory 1990-2000.* Ottawa: Environment Canada.

Executive Resource Group. 2001. *Managing the environment: A review of best practices.* Report prepared for the Ontario Ministry of Environment.

Faiz, A., C. Weaver, and M. Walsh. 1996. *Air pollution from motor vehicles: Standards and technologies for controlling emissions.* Washington, DC: World Bank.

Government of Canada. 2002. *A discussion paper on Canada's contribution to addressing climate change.* Ottawa: Government of Canada.

Grubler, A., N. Nakicenovic, and D. Victor. 1999. Modeling technological change: Implications for the global environment. *Annual Review of Energy and Environment* 24: 545-69.

Harrison, K. 1999. Talking to the donkey: Cooperative approaches to environmental protection. *Journal of Industrial Ecology* 2 (3): 51-72.

Hessing, M., and M. Howlett. 1997. *Canadian natural resource and environmental policy.* Vancouver: UBC Press.

Jaccard, M. 2004. Renewable portfolio standard. In *Encyclopedia of Energy,* ed. C. Cleveland, 5, 413-21. New York: Elsevier.

Jaccard, M., J. Nyboer, and B. Sadownik. 2002. *The cost of climate policy.* Vancouver: UBC Press.

Jaccard, M., N. Rivers, and M. Horne. 2004. *The morning after: Optimal greenhouse gas policies for Canada's Kyoto obligations and beyond.* Ottawa: C.D. Howe Institute.

Jacoby, H., and D. Ellerman. 2004. The safety valve and climate policy. *Energy Policy* 32: 481-91.

Jaffe, A., R. Newell, and R. Stavins. 2002. Environmental policy and technological change. *Environmental and Resource Economics* 22 (1-2): 41-69.

Karamanos, P. 2001. Voluntary environmental agreements: Evolution and definition of a new environmental policy approach. *Journal of Environmental Planning and Management* 44 (1): 67-84.

Kemp, R. 2004. *Zero emission vehicle mandate in California. Misguided policy or example of enlightened leadership?* Working paper, MERIT, University of Maastricht.

Khanna, M. 2001. Non-mandatory approaches to environmental protection. *Journal of Economic Surveys* 15 (3): 291-324.

Kosobud, R., and J. Zimmerman, eds. 1997. *Market based approaches to environmental policy.* New York: Van Nostrand Reinhold.

Loehman, E., and M. Kilgour, eds. 1998. *Designing institutions for environmental and resource management.* Cheltenham, UK: Edward Elgar.

Lutz, M. 1999. *Economics for the common good: Two centuries of social economic thought in the humanistic tradition.* London: Routledge.

Magat, W. 1978. Pollution control and technological advance: A dynamic model of the firm. *Journal of Environmental Economics and Management* 5: 1-25.

National Round Table on the Environment and the Economy (NRTEE). 2002. *Toward a Canadian agenda for ecological fiscal reform: First steps.* Ottawa: NRTEE.

Organisation for Economic Co-operation and Development (OECD). 2003. *Voluntary approaches for environmental policy: Effectiveness, efficiency, and usage in the policy mixes.* Paris: OECD.

Pal, L. 1997. *Beyond policy analysis: Public issue management in turbulent times.* Scarborough, ON: International Thomson Publishing.

Pilkington, A. 1998. The fit and misfit of technological capability: Responses to vehicle emission regulation in the US. *Technology Analysis and Strategic Management* 10 (2): 211-24.

Pizer, W. 1999. Optimal choice of policy instrument and stringency under uncertainty: The case of climate change. *Resource and Energy Economics* 21: 255-87.

Prugh, T., ed. 1995. *Natural capital and human economic survival.* Solomons, MD: International Society for Ecological Economics.

Sagoff, M. 1988. *The economy of the earth, philosophy, law and the environment.* New York: Cambridge University Press.

Sandler, T. 1997. *Global challenges – An approach to environmental, political and economic problems.* New York: Cambridge University Press.

Stavins, R. 1998. What can we learn from the grand policy experiment? Lessons from SO_2 allowance trading. *Journal of Economic Perspectives* 2 (12): 69-88.

Svendsen, G. 1998. *Public choice and environmental regulation: Tradable permit systems in the United States and CO_2 taxation in Europe.* Cheltenham, UK: Edward Elgar.

Svendsen, G., C. Daugbjerg, and A. Pedersen. 2001. Consumers, industrialists and the political economy of green taxation: CO_2 taxation in OECD. *Energy Policy* 29 (6): 489-97.

Tietenberg, T. 1998. Ethical influences on the evolution of the US tradable permit approach to air pollution control. *Ecological Economics* 24: 241-57.

9

Sustainable Production and the Financial Markets: Opportunities to Pursue and Barriers to Overcome

Blair W. Feltmate, Brian A. Schofield, and Ron Yachnin

The term "sustainable development" was recognized globally following publication of the World Commission on Environment and Development report *Our Common Future* (WCED 1987). It defined sustainable development as the now familiar "development that meets the needs of the present without compromising the ability of future generations to meet their own needs" (43). Within the broad framework of this definition, corporate commitment to sustainable development requires that decision makers simultaneously consider the environmental, economic, and social outcomes associated with performance standards and practices. For example, a decision made with a focus on environmental considerations but without forethought to social or economic ramifications is not consistent with the tenets of sustainable development.

During the early 1990s in Canada, companies began to implement sustainable development programs by employing the following protocol (Feltmate and Schofield 1999):

- Develop a company-specific definition of sustainable development.
- Develop company-specific principles of sustainable development that address commitment to environmental, economic, and social best practices.
- Develop a sustainable development policy (effectively the combination of the previous two points).
- Identify practical, meaningful, cost-effective measures that reflect the mandate of the sustainable development policy.

The first step is key: establishing a company-specific definition of sustainable development (Kennan 1995). Typically, any definition of sustainable development shares three common elements (SMAC 1999):

- Key stakeholders are identified.

- A statement of commitment to environmental, economic, and social best practices is presented.
- A statement of commitment to current and future generations is presented.

As sustainable development was pursued by Canadian companies during the early 1990s, notable senior business advocates emerged. For example, Senator J. Trevor Eyton, senior chairman of EdperBrascan Corporation (which in 2005 was renamed Brookfield Asset Management), unreservedly stated that "our corporate group, including the directors and senior officers of Noranda, have concluded that sustainable development is an inevitability. The sooner a corporation reaches that conclusion, the greater its advantage in the future" (personal communication, 1999). Reflecting a similar sentiment, John Mayberry, former chairman and CEO of Dofasco Inc., affirmed that "the framework for all of Dofasco's business strategies is sustainability. This is a belief and commitment that in order to succeed we must be equally committed to the interdependent triple bottom line of sustainable financial management, commitment to our employees and our community and environmental responsibility" (Hillier 2002). At Phelps Dodge Corporation, Doug Yearly, chairman emeritus and chairman of the International Council on Metals and Minerals, was pointed in his assertion that "we need to cement the realization that sustainable development makes hard-headed business sense" (Hasselback 2002).

But why were, and are, so many corporate leaders – and companies like Aliant Telecom, Bank of Montreal, BHP Billiton, Credit Suisse Group, Falconbridge Ltd., IBM, Manulife Financial, Ontario Power Generation, Royal Bank of Canada, Suncor Inc., and TransAlta Utilities Corporation – committing themselves to sustainable development with such strong conviction? In brief, "sustainable development is a value driver/revenue generator" (Kommunalkredit Dexia Asset Management 2004).

Sustainable Development as a Value Driver

Sustainable development can bring value to business through a variety of avenues. For example, benefits can be realized by sustainable development companies through increases in sales to discriminating customers who prefer products that reflect green stewardship, corporate social responsibility, ethical practices, and a growing list of synonymous terms. Further benefits can accrue to sustainable development practitioners as they strive to increase operational efficiency and minimize, for example, the production of waste, which is good for the environment and the bottom line. To illustrate, as Ford Motor Company produces more output (cars) per unit of resource consumed (nickel, copper, aluminum), this generally appeals to shareholders and (some) environmentalists.

The following is a summary of categories of key factors in why sustainable development is good business (Feltmate et al. 2001).

Access to markets/ease of operational start-ups. A company that carries brand as a sustainable development practitioner will generally be welcomed into communities in which it wishes to do business, and will therefore realize the revenue and share price benefits associated with expanded business interest(s). Conversely, if a company is seen as an environmental, economic, or social pariah, it will not be welcomed, and the associated share price impact resulting from such poor public relations, project cancellations, or delays can be substantial. For example, if a mining company builds brand as a "good corporate citizen," it will generally receive accelerated licensing for new operations, whereas if the company is viewed as an environmental outcast, its start-up will generally be opposed, which can be costly. In such cases not only would the company relinquish the $20-40 million typical of a large mining operation's up-front feasibility and exploration costs, but the revenue the mine would have generated during its lifetime would also disappear. Falconbridge Ltd. is a good example of a nickel mining company that has realized the benefits of sustainable development with its mine in Raglan, northern Quebec, where local residents were, and continue to be, enthusiastic supporters of the operation (Falconbridge 2001). Other mining companies, perceived as less committed to sustainable development, have experienced costly delays regarding project start-ups.

Value chain. Customers are increasingly concerned about the harm that corporate practices might cause from environmental, economic, or social perspectives. To retain or gain the business of customers, which will ultimately affect share price, companies are increasingly adopting recognized business practices that demonstrate corporate citizenship. For example, the International Chamber of Commerce (ICC) and the International Organization for Standardization (ISO) have developed objective criteria to assess environmental product claims utilizing http://www.14000registry.com, which is an online market registry for companies to post their adoption of ISO 14001 and to provide a link to their website.

Media/activist pressures. Organizations such as Natural Step, Greenpeace, American Rivers, Sierra Club, and Friends of the Earth can affect public perceptions of business, thus affecting customers' buying practices, product switching, operational start-ups, and ultimately share price. To gain or retain the support of these organizations, and, at minimum, to not suffer their assaults, a company serves itself well to demonstrate its commitment to environmental, economic, and social stewardship, and to engage in a dialogue with these organizations to proactively identify potential omissions in practices.

Lower bank loan rates. Most major banks employ senior environmental managers to assess the cumulative environmental risk associated with lending capital for mortgage holdings, land acquisitions, and so on. Interest rates are adjusted to reflect risk, and companies that are positioned as sustainable development practitioners are perceived as presenting less risk; accordingly, the cost of borrowed capital is reduced and the company retains cash, which will generally have a positive impact on share price. Also, as banks are increasingly concerned with issues of lender liability, the success of a company to gain a loan, at any cost, is affected by the environmental, economic, and social practices of the company. According to Sandra Odendahl, Senior Manager, Environmental Risk, at the Royal Bank of Canada (personal communication, 2002):

> The Royal Bank integrates a comprehensive environmental review process when assessing loan applications. While we may approve financing for a poor environmental performer with an appropriate corrective action program, typically companies that do not have a solid track record in environmental management present a higher risk to the bank. As a result, these companies are less likely to receive approval than those able to demonstrate sound environmental performance. If they do receive approval, their cost of funds will probably be higher.

Lower insurance premiums. Although many major companies do not carry environmental insurance (many companies are self-insured), insurance companies are including sustainable development and environmental risk in their underwriting process. Companies that are sustainable development practitioners and that are not self-insured will generally receive discounted premiums, which translate into savings that can have a positive impact on share price. In 1998, American International Group, the largest underwriter of industrial insurance in the United States, took the first steps to factor measures of corporate sustainable development into its underwriting process (http://www.aigonline.com).

Increased eco-efficiency of operations. Eco-efficiency is born of a contraction of *eco*logical and *eco*nomic efficiency, and it advocates *doing more with less*. For example, an eco-efficient company will reduce energy input, material requirements, and waste production per unit of productivity. In turn, the company will retain more cash for alternative applications that, if applied effectively, can have a positive impact on share price.

Due diligence regarding partnerships and acquisitions. Due diligence requires that the sustainable development performance of partners or acquired companies be assessed. Engaging in a relationship with a company that has a negative reputation can result in potential liabilities. If a company carries the brand of a sustainable development practitioner, there is a higher probability that it will be engaged as a partner and derive associated benefits. Similarly, if a company is to be acquired, a positive brand as a sustainable development practitioner can attach a premium to its share price.

Facilitation of divestiture. For reasons just described, companies with a positive record of sustainable development performance will generally realize a higher valuation for shareholders upon sale. Due diligence requires the assessment of sustainable development, prior to divestiture, as a factor for inclusion in the valuation process.

Self-regulation. When industry and government combine expertise regarding the application of sustainable development best practices, practical and cost-effective self-regulatory programs and/or legislation will often emerge. History suggests that sustainable development programs arising from collaboration between industry and government are generally preferable from a share price perspective, compared with programs developed through independent efforts. To illustrate, the Voluntary Challenge and Registry (VCR), sponsored by the Canadian Industry Program for Energy Conservation and Natural Resources Canada, was a program that challenged Canadian organizations to voluntarily limit or reduce greenhouse gas emissions to 1990 levels. Many companies – such as Canadian Natural Resources Ltd., EnCana Corporation, EPCOR Utilities, Imperial Oil Ltd., and Petro-Canada – that had voluntarily elected to participate in the VCR simultaneously opposed the Kyoto Protocol, which would drive these organizations to reduce greenhouse gas emissions to 6 percent below 1990 levels by 2008-2012. These companies considered the projected costs resulting from the Kyoto accord to be untenable (Houlder 2002), whereas the self-regulatory approach of the VCR was viewed favourably.

Employee satisfaction/retention. Companies that are practitioners of sustainable development report that most employees welcome challenges associated with environmental, economic, and social stewardship. Accordingly, employee job satisfaction scores generally increase within one to three years following the initiation of sustainable development programs; employee productivity increases; and the service time of employees with a company increases (thus lowering start-up training costs). All of these factors generally have a positive impact on share price appreciation. For example, Gaynor McEwan, former Program Manager of Diversity and Workplace Programs for IBM Canada, explains that "efforts to embrace employee diversity [e.g., based on gender, culture, lifestyle] directly benefit IBM through many avenues, such as enhanced employee job satisfaction, lower workplace turnover and increased productivity" (personal communication, 2002).

Inclusion in sustainable development and environmental funds. As indicated in Table 9.1, a large and growing number of mutual and institutional funds apply sustainable development or social screens to portfolio construction. Corporate sustainable development programs can facilitate a company's inclusion in these portfolios, resulting in a positive impact on share price.

Besides the direct drivers outlined above, sustainable development can function as a proxy for identifying companies with superior quality of management. By definition, sustainable development companies are interdisciplinary in their approach to business, which reflects a senior management team that thinks ahead of the curve (*Economist* 2004; Griss 2000). Accordingly, firms pursuing sustainable production strategies would probably make superior decisions with respect to such key business issues as product and production line innovation, new financing options, exploitation of new market opportunities, executive hiring, and so on, and so these companies present preferred investment options (Ehrbar 1998). Jones Heward Investment Counsel Inc. and Bank of Montreal chairman Barry Cooper summarized the relational benefit of sustainable development investing very well: "Sustainable development offers an objective means to identify companies with superior quality of management that thinks long-term" (personal communication, 2002).

Sustainable Development Style Investing

The inclusion of sustainable development measures of performance applied to investment management is reflected in both retail and institutional funds and private wealth portfolios that utilize sustainable development to guide investment decisions (Feltmate et al. 2001). The University of Basel identified 260 "sustainable development style" retail funds that were available globally as of August 2004. The distribution of these funds by country is presented in Table 9.1.

Sustainable development style investing is most prominent in the United States, where socially responsible investing (SRI) grew from $1.2 trillion in 1997 to $2.2 trillion in 1999. In other words, SRI accounts for over 13 percent

Table 9.1

Number of sustainable development style funds per country (as of August 2004)

Country	Number of funds	Country	Number of funds
United States	98	Japan	7
United Kingdom	46	Australia	6
Sweden	30	South Africa	4
Canada	29	Belgium	3
Switzerland	15	Finland	2
France	9	Norway	2
Germany	8	Mexico	1

Source: http://www.sustainablevalue.com.

of the US$16.3 trillion in investment assets under professional management in the United States (Social Investment Forum 2000). In Canada, SRI accounts for 3.2 percent, or Cdn$50 billion, of assets held in mutual, institutional, and socially screened labour funds (http://www.socialinvestment.ca). In North America, growth in SRI assets under administration is 40 percent per year, compared with 15 percent per year across the broader markets.

SRI funds often employ *negative or exclusionary screens* to identify companies appropriate for inclusion in portfolios. Negative screens select against investment in products or processes that are, in the estimation of the portfolio manager, deemed not to support the public good (SIF 2000). This has meant that in the US, 96 percent of all SRI funds have rejected holdings in tobacco companies, 86 percent exclude gambling, 83 percent exclude alcohol, and 81 percent reject weapons manufacturers. Alternatively, in the late 1990s, funds that employed *positive or inclusionary screens* began to emerge, with a focus on environmental, economic, and social practices – collectively, sustainable development. The next section explores some of the key characteristics of the most prominent negatively and positively screened European and North American funds.

Prominent Funds and Indices

Indices and funds referenced below are those that are generally well recognized in the countries where they are managed. Within particular fund families, the focus is on those funds with the longest track records.

Dow Jones Sustainability Group Index

The Dow Jones Sustainability Group Index (DJSGI) (http://indexes.dowjones.com) is jointly administered and managed by US-based Dow Jones Indices and Swiss-based Sustainable Asset Management (SAM) (http://www.samswiss.ch). As of 2004, the DJSGI consisted of over 300 companies deemed to be "meaningful practitioners of sustainable development," selected from 2,000 stocks with the largest market capitalization in the Dow Jones Global Index (DJGI). The DJSGI represents thirty-four countries and sixty industry sectors. At the end of 2004, the market capitalization of the DJSGI exceeded US$6.2 trillion (Dow Jones Sustainability Indexes 2005). The DJSGI methodology facilitates sustainability-specific indexing and employs negative screens – for example, ex-tobacco, ex-alcohol, and ex-gambling on one global, three regional, and one country scales (i.e., World, Asia/Pacific, Europe, North America, US). Based on a comparison of eleven years of total returns, the DJSGI outperformed the Morgan Stanley Capital International (MSCI) World Index by approximately 48 percent (see Figure 9.1).

Domini Social Equity Fund

Launched in 1991, the Domini Social Equity Fund describes itself as the

Figure 9.1

Performance of the Dow Jones Sustainability Group Index (DJSGI) versus the Morgan Stanley Capital International (MSCI) World Index, 1993-2004

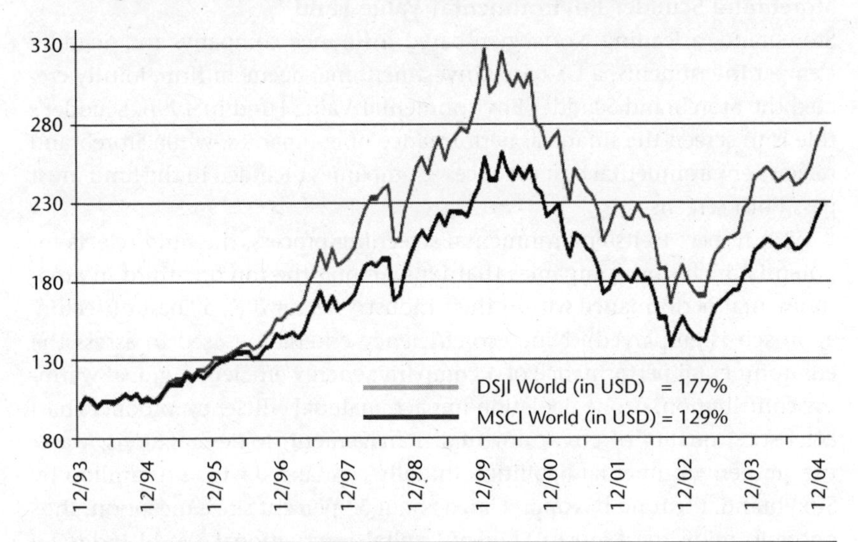

Source: Dow Jones Sustainability Group Index 2005.

first socially responsible index fund in the United States (Domini Social Equity Fund 2000). The fund employs negative screening and holds 400 stocks that comprise the Domini 400 Social Index. The index does not invest in companies that manufacture tobacco, alcohol, or nuclear power; supply services to gambling operations; or derive more than 2 percent of gross revenues from the production of military weapons. It selects for companies that demonstrate solid environmental performance.

The ten-year annualized return (1991-2004) of the fund was 10.4 percent, which is effectively identical to the return of the S&P 500 (10.9 percent) over the same period. This return is impressive considering that most managed funds in the US substantially underperform the S&P 500.

EcoValue 21™

Innovest Strategic Value Advisors founded EcoValue 21™ in 1998. This system uses up to sixty environmental criteria to develop its EcoValue 21™ ratings of AAA (outperform) to CCC (underperform) on over a thousand domestic and international equities. Depending upon the industry sector, companies with above average EcoValue 21™ ratings have consistently outperformed lower-rated companies, on a share price appreciation basis, by 3-25 percent per year based upon back-tested analyses. Thus, EcoValue 21™ is a tool to "identify hidden value and risk potential among companies using evaluation techniques often ignored by mainstream Wall Street analysts" (EcoValue 21™ 2000).

Storebrand Scudder Environmental Value Fund

Storebrand, a leading Norwegian-based insurance company, and Scudder Kemper Investments, a US-based investment management firm, jointly created the Storebrand Scudder Environmental Value Fund in 1996. Scudder's role is to screen the financial performance of companies, while Storebrand reviews environmental best practices. Companies included in the fund must pass both screens.

With respect to its environmental screening process, the fund selects for potential inclusion companies that rank among the top one-third in environmental performance within their industry sector (i.e., a "best of sector" approach is employed). Nine eco-efficiency criteria are used to assess the environmental performance of a company: energy efficiency, global warming contribution, ozone depletion impact, material efficiency, product characteristics, quality of environmental management, toxic emissions, water use, and environmental liabilities. Initially capitalized with $10 million by Storebrand, the fund has appreciated about 51 percent since inception, thus outperforming the Morgan Stanley Capital International World Index by approximately 8 percent (Deutsch 1998).

Ethical Funds®

Ethical Funds®, located in Canada, manages portfolios based on the application of "ethical screens" developed by Michael Jantzi Research Associates (MJRA) and Kinder, Lydenberg, Domini & Co. Inc. (KLD) of Boston (http://www.ethicalfunds.ca). Ethical Funds® excludes portfolio holdings in companies associated with tobacco, the nuclear industry, or military production. Relative to positive screens, the funds seek investment in companies with progressive environmental, industrial, and community relations practices, and that adopt and promote human rights standards. Five Ethical Funds® with a ten-year track record are profiled in Table 9.2. Of these, the Money Market Fund, Balanced Fund, Growth Fund, and North American Equity Fund underperformed group average and index returns. The Income Fund effectively matched the group average, yet it underperformed the 10-year index by 1.34 percent. With little exception, therefore, with respect to Ethical Funds® the "ethical" screening process appears to produce portfolios that underperform benchmarks.

Sustainable Development Fund® and Sustainable Development Index (SDI®)

The Sustainable Development Fund® was launched in Canada in August 2001, by Jones Heward Investment Counsel Inc., a money management

Table 9.2

Ten-year compound rates of return for Ethical Funds® compared with group average and ten-year index returns (for period ending 31 December 2004)

Fund	Total assets (million $)	10-year Ethical Fund® return (%)	10-year group average returns (%)	10-year index returns (%)
Credential® Money Market Fund	125.8	3.11	3.68	4.29
Ethical® Income Fund	223.7	7.69	7.51	9.03
Ethical® Balanced Fund	407.1	6.65	8.05	9.58
Ethical® Growth Fund	435.8	6.68	9.40	10.08
Ethical® North American Equity Fund	117.8	6.46	6.95	10.37

Source: http://www.globefund.ca.

firm wholly owned by the Bank of Montreal (Papmehl 2002). The fund consisted of approximately forty-five large capitalization equity securities that demonstrate a quantifiable commitment to sustainable development.

The Sustainable Development Fund® identifies companies that meet its predefined sustainable development standard using its Sustainable Development Index (SDI®). The SDI® focuses on 80-160 industry-specific environmental, economic, and social measures to assess a company's performance. The number of measures in the analysis varies according to industry sector (e.g., the chemical sector analysis focuses on 155 measures, whereas the financial sector analysis focuses on 60). Industry specificity is necessary due to idiosyncratic differences between the operations of banks, mining companies, hotel chains, chemical companies, and so on. Evaluative information regarding the practices of a company is based upon random site visits, interviews (i.e., with vice presidents, environmental health and safety (EHS) managers, operating officers, etc.), questionnaires, and reviews of sustainable development/EHS reports. Companies that receive an SDI® score greater than 70 percent (out of a theoretical maximum of 100 percent) are eligible for inclusion in the Sustainable Development Fund®.

The fund utilizes a positive or inclusionary approach for portfolio construction. During the period August 2001 to March 2003, it realized a return of 1.1 percent, whereas the S&P/TSX lost 10.2 percent during the same period, yielding a total differential of 13.3 percent upside performance of the fund compared with its benchmark. Additionally, the fund consistently outperformed the RT Median Equity Managers Return (see Figure 9.2).

With few exceptions, sustainable development style portfolio management suggests that corporations that practice environmental, economic, and social stewardship tend to realize share price appreciation that matches

Figure 9.2

Performance of the Sustainable Development Fund® versus the S&P/TSX and RT Median Equity Managers Return, August 2001 to March 2003

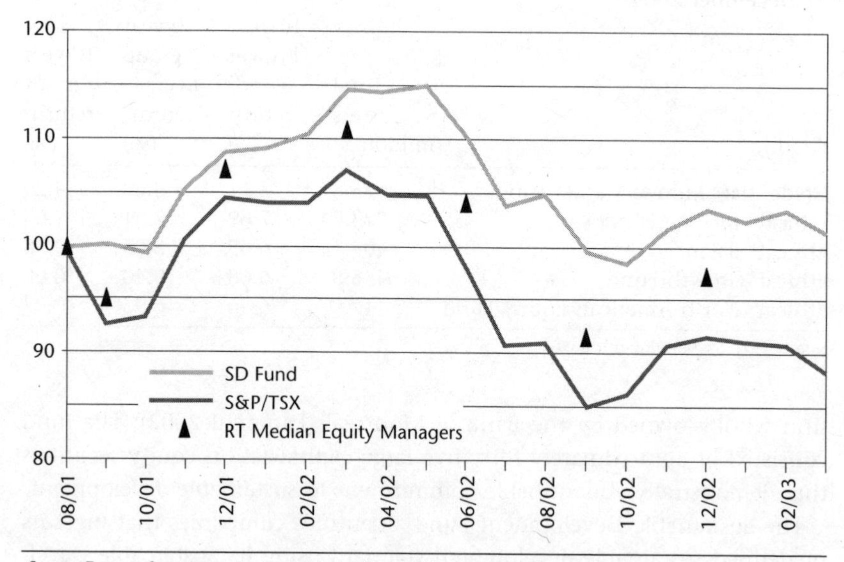

Source: Personal communication, Jones Heward Investment Counsel Inc.

or outperforms market benchmarks. As discussed below, a growing body of research explains why these returns may be expected to continue.

Review of Shareholder Value Studies

Evidence to support the claim that corporate sustainable development practice has a positive impact on share price is growing. For two reasons, however, absolute proof of cause and effect is difficult to establish (Schaltegger and Figge 1997). First, universally agreed-upon measures used to identify a company as a "sustainable development" practitioner do not exist (Feltmate et al. 2001). Without such measures, there will always be a question as to whether a particular company's identification as a sustainable development practitioner is justified. Accordingly, attributing the share price performance of a company to sustainable development performance can be treated only as "evidence of a relationship." Second, sustainable development correlates with a multitude of corporate quality-of-management issues that can impact share price. Accordingly, it can be extremely difficult to separate the influence of sustainable development initiatives from a multiplicity of other factors that also influence share price. For example, when a sustainable development company makes prudent decisions with respect to product innovation, financing options, new markets opportunities, or strategic alliances,

Figure 9.3

Key messages of studies focusing on the relationship between sustainable development and shareholder value

Title of study	Key message
Sustainability Pays Off (Kommunalkredit Dexia Asset Management 2004)	"Companies favouring the concept of sustainability outperform the broad market."
Meta-Analysis Reveals Clear Linkage Between CSP and CFP (Orlitzky 2004)	"Corporate [SD] performance and financial performance are generally positively correlated across a wide variety of industry and study contexts."
2002 Sustainability Survey Report (Savitz et al. 2002)	"Companies that fail to become sustainable – that ignore the risks associated with ethics, governance and the 'triple bottom line' of economic, environmental and social issues – are courting disaster. In today's world of immense and instant market reaction, an action or inaction that undermines the integrity, ethics or reputation of a company can lead to immediate and dire financial consequences."
Single Bottom Line Sustainability (Gilding et al. 2002)	"Companies that produce more social and environmental value also produce more financial value."
From Economics to Sustainomics (Hendrik et al. 2002)	"Companies whose activities no longer find social and environmental acceptance are sanctioned in the sales and procurement markets. Products can no longer be sold, workers opt for other employers, and financiers demand an additional risk premium to protect themselves against reputational risks."
Greening Your Pension Fund (Williams and George 2001)	"Some [pension] plan sponsors already state that social responsibility is an organizational value. This makes socially responsible investing (SRI) a logical consideration for these organizations."
Sustainable Development: The Next Step in Evolution for Socially Responsible Investing (Feltmate and Schofield 2001)	"Recognizing that sustainable development is a value driver/revenue generator for business, sustainable development has entered the capital markets as a mainstream portfolio management tool ... Sustainable development portfolio management focuses foremost on enhancing performance."

►

◂ *Figure 9.3*

Title of study	Key message
Big Business Puts Environment in Its Big Picture (Bent 2000)	"Business analysts and decision-makers from France, Germany and the United Kingdom concluded that corporate environmental responsibility is a *need to do*. Leading companies are recognizing that creating value for both society and shareholders are mutually reinforcing objectives."
Ethical Investment: Morality Plays (*Economist* 2000)	"In the long run, there need not be a trade-off between profits and ethics."
For Better or Worse: Can Corporate Commitment to Sustainability Survive a Marriage? (Griss 2000)	"Evidence is mounting that companies with good performance on environment and sustainability issues tend to also do well in other areas."
Green Investment: Ecological Wake-up Call for Fund Managers (www.FT.com 1999)	"Mainstream companies with good environmental records have tended to perform well. Studies, mostly from the US, have found a significant correlation between the environmental credentials of companies and share price performance."
Shades of Pink and Green: Financial Markets Have Yet to Get Serious about the Environment (Spencer-Cooke 1999)	"An analysis of the history of 300 companies in the Fortune 500 suggests that environmentally minded companies outperform competitors by around 2-3 percent."
Social Funds a Novelty No More (Prial 1999)	"Vanguard Group and TIAA-CREF both plan to open socially responsible funds. When names like this decide to join a trend, it's no longer a trend – it's the mainstream." *Note:* Vanguard is the second largest mutual fund company in the US, with US$521 billion in assets. TIAA (Teachers Insurance and Annuity Fund) and CREF (College Retirement Equities Fund) have combined assets of US$250 billion.
Stern Stewart's EVA (Ehrbar 1998)	"Companies that stressed a strong commitment to ethics had an average MVA (Market Value Added – a measure of shareholder wealth creation) of $16.8 billion, versus an average MVA of $11.1 billion for those with some commitment and $5.7 billion for those with no mention of ethics [in annual reports]. Value- creating companies recognize the importance of ethical behaviour."

▸

◄ *Figure 9.3*

Title of study	Key message
Investor Attitudes toward the Value of Corporate Environmentalism: New Survey Findings (Soyka and Feldman 1998)	"Our findings suggest that money managers believe that investments in EH&S improvements can create incremental value, and that if this value is convincingly demonstrated, they are willing to pay a premium for the equity issued by the company."
Ethics, Environment and Enterprise (Strong 1998)	"Individuals have come to recognize that significant amounts of capital can be earned by investing in environmentally and ethically friendly companies."
Uncovering Value: Integrating Environmental and Financial Performance (Riggs 1998)	"Like leading edge companies, analysts and investors who seek and use information about the business results of environmental linkages will have an advantage over their peers. Although the context may be new, the questions deal with familiar subjects: quality of management, risk exposure, brand image and reputation, overall operating efficiency, growth and market access."
Wake-up Call for Fiduciaries: Eco-efficiency Drives Shareholder Value (Kiernan and Martin 1998)	"Eco-efficiency can be defined as the capacity to create greater shareholder value with levels of resource inputs and environmental risk less than one's corporate competitors."
Green Shareholder Value, Hype or Hit? Sustainable Enterprise Perspectives (Reed 1998)	"While the end game is not in sight, the trends are clear enough that corporate managers would be foolhardy not to take action now, positioning to take advantage of greater attention to their environmental performance by shareholders."
Environmental Shareholder Value (Schaltegger and Figge 1997)	"The study demonstrates that environmental protection measures which enhance enterprise value ... increase sales, raise margins, protect the flow of finance and increase the company's value over the long term."
Environmental Performance and Shareholder Value (Blumberg et al. 1997)	"The financial community can improve the quality of its decision-making by integrating companies' environmental performances into its analyses. Environmental issues can drive financial performance through traditional elements of financial evaluation such as robust and forward looking strategies, operational fitness of the company, product quality and markets, and company reputation. They manifest

►

◄ *Figure 9.3*

Title of study	Key message
	themselves through the elements of eco-efficiency, the recognition and management of risks, the quality of business management and the identification of new business opportunities. The quality of a company's environmental management provides the outside world with a good indicator of the overall quality of its business management."
A Resource-Based Perspective on Corporate Environmental Performance and Profitability (Russo and Fouts 1997)	"Results indicate that it pays to be green and that this relationship strengthens with industry growth."
Does Improving a Firm's Environmental Management System and Environmental Performance Result in a Higher Stock Price? (Feldman et al. 1997)	"Our results show that firms will increase shareholder value if they make environmental investments that go beyond strict regulatory compliance. Findings suggest that investments in environmental management and improved performance can be justified, in many cases, on purely financial grounds."

the consequences of these decisions will also impact share price. Isolating the impact on share price attributable to sustainable development performance compared with other confounding variables is a task requiring at least ten years of data collection and multiple regression followed by stepwise regression analysis (Bhattacharyya and Johnson 1986).

As a complement to direct measures of causality, studies referenced in Figure 9.3 provide insight into perceived relationships between sustainable development and shareholder value and share price.

Emerging Sustainable Development Financial Drivers

Internationally, changes in institutional investment policy are positioning sustainable development further in mainstream investing (Feltmate et al. 2001). For example, in the United Kingdom, it is the position of the government that financial institutions are, and should be, important instruments to facilitate the application of sustainable development within business (Borremans 2000). Accordingly, through amendments to UK pension fund regulations that took effect in July 2000, all funds are now required to state whether they take account of the environmental, social, and ethical impacts of their investments (Carlstrand 2000). The amendments do not require any fund to change its policy, but "its clear purpose is to shine a fiery spotlight on their practices. Socially Responsible Investing fans believe that many funds will now look seriously at ethical investing for the first time, if only to avoid appearing callous" (*Economist* 2000).

The ramifications of the amended UK pension fund requirements are substantial and potentially far-reaching. For example, a survey (May-June 2000) of the twenty-five largest pension funds in the UK by the research organization Environmental Resources Management (2000), revealed that twenty-one of these funds intended to implement socially responsible investing principles into their statement of investment policies and guidelines. These funds collectively invest £120-150 billion in UK equities, representing 7-10 percent of UK stocks.

The key message of this survey is that pension fund managers and financial analysts will increasingly look for companies whose shareholder value is enhanced, or at least protected, by prudent management of environmental and social risks. Pressure is now on companies to (1) adopt best-practice policies for managing environmental and social risks, (2) demonstrate the business case for proactively managing environmental and social risks, and (3) adequately communicate this information to investors (Borremans 2000).

Perhaps the most meaningful of the UK pension fund amendments is the fact that sustainable development may now be factored into the investment decision-making process without the concern of violating fiduciary responsibility (i.e., the legal requirement that a pension fund manager's primary mandate is to maximize returns and not engage any investment criteria that might otherwise compromise returns) (Feltmate et al. 2001). In fact, recognizing the value creation and lower risk associated with sustainable development practices, it is now arguable that investment decisions made without assessing the environmental, economic, and social practices of companies may stand in violation of fiduciary responsibility. The following example illustrates how poor environmental performance can negatively impact share price.

In April 1998, the Toronto-based company Boliden Ltd. (45 percent owned by Trelleborg AB of Sweden) experienced a tailings pond rupture at its zinc mine in Spain, releasing a flow of toxic waste that contaminated nearly 20 square kilometres of downstream farmland (Gibson 1998). The day after the accident, the share price of Boliden dropped from Cdn$11.90 to $10.45. Approximately two and a half years later, the company's shares traded at $1.05. In the estimation of many analysts, this sharp decline in share price was substantially influenced by the April 1998 accident.

Besides amendments to the UK pension fund requirements, two meaningful events have taken place in the UK capital markets (Feltmate et al. 2001).

Notably, the Turnbull Report is "ringing alarm bells for companies listed on the London Stock Exchange" (Davidson 2000). Produced by the Internal Control Working Group with the support of the government and the Institute of Chartered Accountants, the report requires companies listed on the London Stock Exchange to itemize and account for all their "risks" – financial, environmental, social, ethical – and report on them at their year-end, starting December 2000. Fear remains that in the longer term it may be difficult for companies to insure themselves, and that company directors could be open to lawsuits.

Furthermore, the stock market index Financial Times Stock Exchange (FTSE) announced the launch of the FTSE4Good Index series, which will provide tradable benchmark indices for the rapidly growing area of socially responsible investing (SRI) (Merrifield 2001). The index will take account of social, environmental, and ethical issues pertaining to corporate best practices. FTSE CEO Mark Makepeace stated that his decision to launch the index series was partly influenced by his commitment to "encourage companies to adopt socially responsible principles."

UK pension fund amendments, the Turnbull Report, and the launch of the FTSE4Good Index series may positively influence the North American capital markets to more readily accept sustainable development as an institutional investment criterion.

In Canada, there is no overriding legal consensus as to whether a sustainable development mandate within institutional investing meets fiduciary responsibility; conversely, there is no legal consensus suggesting that it does not. At the provincial level, however, there has been some guidance. For example, the Financial Services Commission of Ontario has indicated that "ethical investing is permitted in pension plans, but the Statement of Investment Policies & Guidelines must state this position and set out the criteria for investments. The members of the plan should be notified of this position" (Financial Services Commission of Ontario 1992). Similarly, the Manitoba Trustee Act (1999) allowed the use of non-financial criteria, including environmental and ethical evaluations of companies, to be used within pension fund investment decision making.

In the United States, the Department of Labor's Office of Regulations and Interpretations issued a May 1998 letter that helped settle a lingering question about whether a socially responsible mutual fund could be included in retirement plans that qualify under Section 404c of the Employment Retirement Income Security Act (ERISA) (SIF 1999). The letter clarified that investments such as socially screened mutual funds could be included in an ERISA-qualified retirement plan, as long as the fiduciary determines that the mutual fund is expected to provide an investment return similar to alternative investments having similar risk characteristics. This clarification is helping to spur the use of socially screened funds in retirement plans. It has also helped provide a greater level of comfort for trustees and fiduciaries to utilize sustainable development funds for other types of institutional investments. Increasing recognition of sustainable development by the financial communities of the UK, US, and Canada may presage a growing recognition of sustainable development by the capital markets globally.

Impact of Sustainable Development on Valuation: Key Issues for Consideration

For issuers of equity and debt securities in either the public or private markets, pricing is established through communication to and valuation by the financial community. The chief financial officer (CFO) is the primary conduit through whom financial information is transferred to the broader financial community. It is therefore essential that the CFO be fully conversant in the language of sustainable development. As obvious as this logic may be, in many cases CFOs and other senior officers do not recognize sustainable development as a value driver, and they have difficulty articulating its financial merits.

Recognizing that the investor is the ultimate arbiter of equity or debt price, it is critical that corporations communicate the merits of all aspects of a business's operations. As described throughout this chapter, the financial community and investors are *beginning* to be influenced by a company's sustainable development activities, and the valuation of a company and the pricing of securities are reflecting this influence. To appreciate the nature of this relationship, a brief review of business valuation factors is necessary.

In general, business valuations in the capital markets are based upon three factors: (1) the magnitude of all future cash flows, (2) the timing of these cash flows, and (3) the risk associated with receiving these cash flows. Large institutional pools of capital drive the share price of public companies because money managers are constantly considering the flow of information about the company and its industry and how this information affects these three valuation factors. Similarly, credit rating agencies, such as the Dominion Bond Rating Service (DBRS) and Canada Bond Rating Service (CBRS), assess credit risk based on the current financial capacity of the borrower and the likelihood of that capacity improving or eroding over time, based predominantly on these same three factors.

All valuation methods employed by research analysts and money managers are variations of this basic "cash flow" model. Rules of thumb have also been established that enable comparisons between companies within business sectors, and in some cases between sectors, to estimate the relative attractiveness of investing in one company versus another.

For example, with respect to a utility company, five valuation methods that would traditionally be employed by the financial community are: (1) Price/Earnings (P/E) multiples compared with industry averages, (2) Price/

Cash Flow (P/CF) multiples compared with industry averages, (3) Price/Book Value (P/BV) multiples compared with industry averages, (4) Discounted Cash Flow (DCF) analysis, and (5) Economic Value Added (EVA). The first three methods are relative measures of value, as indicated. The DCF and EVA methods are absolute valuation measures and are calculated directly from the financial forecasts of the company.

From a sustainable development perspective, and as a complement to the five factors, it is essential to collect and process additional "non-traditional" (e.g., sustainable development) information to fully appreciate the company's valuation in the equity and debt markets. To do so requires three steps: (1) gather sustainable development performance information about the company from all sources, (2) transform the information into financial valuation parameters, and (3) communicate the information to the financial community in financial or valuation terms. A brief description of each step follows.

Gathering Sustainable Development Information

The documentation of a company's sustainable development practices should be automatic and performed in conjunction with the financial tracking of information regarding projects and operations. Accordingly, a database that can be used to relate sustainable development practices to financial valuations should be established.

Transforming Information into Financial Valuation Parameters

Information regarding the benefits that sustainable development practices bring to business operations must be translated into financial impact. For example, how does the practice of sustainable development facilitate regulatory approval for a plant expansion, secure sales to a new customer, or reduce waste associated with daily operations? Most of the benefits realized through sustainable development practices can be directly quantified in terms of financial impact in one or more of the following categories: (1) increasing revenue, (2) reducing costs, and (3) lowering risk. Each of these financial benefits is an element in the valuation methods listed above, directly impacting the magnitude of the valuation.

Sustainable development practices within a company can thus be directly valued, providing the immediate valuation impact of such activities for the organization, on a case-by-case basis or in aggregate. Evaluating projects based on their financial benefits using DCF or another method has traditionally been done as part of the capital budgeting process and, admittedly, can be a difficult exercise involving numerous estimates. Extending this well-developed technique to sustainable development endeavours is more complex, given that typically less quantitative information is known regarding the specific magnitude and timing of the cash flow impacts of sustainable development practices. Nonetheless, estimating the financial impact that sustainable development practices have on operations on a business segment or project basis provides valuable insights both for internal management and for the financial community.

Communicating the Information

To realize the full incremental value associated with sustainable development practices, the CEO and CFO of a company should communicate this information to the other members of the financial community. At this early stage of building sustainable development factors into valuation models, it is critical that the most senior company representatives be seen as the providers of this information. The results of the valuation analyses should be disseminated to the financial community on both a business segment or project basis, and also on an overall company basis.

For example, if energy-efficient programs have increased cash flow in a particular major operation by 3 percent, this should be articulated. Furthermore, if all sustainable development activities have increased cash flow by 5 percent and earnings by 7 percent, the company should isolate these benefits and, as described above, estimate the value impact using an appropriate valuation method. It sends a powerful message, for example, to announce that sustainable development has contributed $1.50 (+7.5 percent) to a $20 stock price.

Once the sustainable development activities can be described in financial terms, company spokespersons should take every opportunity to discuss these benefits in the usual forums for financial discussions, including annual and quarterly reports, press releases, quarterly analyst conference calls, and media interviews. Particularly successful sustainable development results can be profiled in newspaper articles and general business magazines or trade journals. Priority should be given to presentations to institutional money management consultants (e.g., Brockhouse Cooper, Frank Russell Canada Ltd., Towers Perrin, William M. Mercer Ltd.) and financial organizations such as the Toronto Society of Financial Analysts, the Treasury Managers Association of Canada, or the Pension Investment Association of Canada.

Conclusion

As evidenced throughout this chapter, sustainable development is emerging as a value driver that is positively related to share price appreciation and that has utility within investment decision making. Nonetheless, recognizing that corporate commitment to sustainable development is often viewed by both retail and institutional investors as the "cost of doing business" (Feltmate and Schofield 1999), it is important that "value creators" (e.g., president, CFO, treasurer, chief information officer, vice presidents, senior managers, and employees) within the financial community more aggressively communicate the value added by corporate stewardship to "value assessors" (e.g., research analysts, brokers, pension and mutual fund managers, and institutional consultants).

Besides the participation of the financial community, greater involvement by government, non-governmental organizations, unions, professional associations, and special interest groups in propagating the message that "sustainable development is a value driver" would further move sustainable development into the mainstream of business practices and investment decision making. As the evidence linking sustainable development to value creation continues to grow, governments should be able to draw a parallel between sustainable development and good governance.

Concerned individuals should assess for themselves how sustainable development factors into personal investment decision making through, for example, mutual fund and defined contribution pension plan investments. Also, at the level of pension plans, pension fund managers should be asked how sustainable development factors into – if it does at all – portfolio construction.

In sum, good news generated by sustainable development is beginning to be appreciated in the capital markets, both in Canada and internationally. The challenge now is to effectively meld the interests of two traditionally non-interacting parties – the sustainable development community and the capital markets – into a mutually reinforcing framework of mainstream investing.

References

Bent, N. 2000. *Big business puts environment in its big picture.* ENN News: http://enn.com.

Bhattacharyya, G., and R. Johnson. 1986. *Statistical concepts and methods.* Toronto: John Wiley and Sons.

Blumberg, J., A. Korsvold, and G. Blum. 1997. *Environmental performance and shareholder value.* Conches-Geneva: World Business Council for Sustainable Development (WBCSD).

Borremans, E. 2000. *Survey shows majority of top 25 UK pension funds will implement socially responsible investment.* http://www.erm.com.

Carlstrand, V. 2000. Ethical injection: New pension fund legislation in the UK will pump up socially responsible investment. *Tomorrow* 10 (3): 52.

Davidson, H. 2000. Samaritan Inc.: More and more customers – and maybe even government – are demanding that business become socially responsible. *Profit* 19 (3): 24-28.

Deutsch, C. 1998. For Wall Street, increasing evidence that green begets green. *New York Times,* 19 July.

Domini Social Equity Fund. 2000. Domini Social Equity Fund homepage. http://www.domini.com.

Dow Jones Sustainability Group Index. 2005. Dow Jones Sustainability Group Index homepage. http://indexes.dowjones.com.

Dow Jones Sustainability Indexes. 2005. Key facts – DJSI world. http://www.sustainability-index.com/htmle/djsi_world/keyfacts.html.

Economist. 2000. Ethical investment: Morality plays. *The Economist,* 8 July.

–. 2004. Corporate storytelling. *The Economist,* 6 November.

EcoValue 21™. 2000. EcoValue 21™ homepage. http://www.innovestgroup.com.
Ehrbar, A. 1998. *Stern Stewart's EVA: Economic value added.* New York: John Wiley & Sons.
Environmental Resources Management. 2000. Pension fund commitment to social investment. http://www.erm.com.
Falconbridge Ltd. 2001. *2001 Sustainable Development Report.* Toronto: Falconbridge.
Feldman, S., P. Soyka, and P. Ameer. 1997. *Does improving a firm's environmental management system and environmental performance result in a higher stock price?* Fairfax, VA: ICF Kaiser International.
Feltmate, B., and B. Schofield. 1999. Elevating shares and saving the planet. *CMA Management* 73 (6): 20-24.
Feltmate, B., and B. Schofield. 2001. Sustainable development: The next step in evolution for socially responsible investing. *University Manager* (Winter): 37-38.
Feltmate, B., B. Schofield, and R. Yachnin. 2001. *Sustainable development, value creation and the capital markets.* Ottawa: Conference Board of Canada.
Financial Services Commission of Ontario. 1992. Ethical investments: Investment of pension funds. *The Superintendent of Pensions,* index no. 1400-350.
Financial Times. 1999. Green investment: Ecological wake-up call for fund managers. http://www.FT.com.
Gibson, A. 1998. Boliden stock battered after toxic spill in Spain. *Globe and Mail,* 28 April, B19.
Gilding, P., M. Hogarth, and D. Reed. 2002. *Single bottom line sustainability.* Sydney: Ecos Corporation.
Griss, P. 2000. For better or worse: Can corporate commitment to sustainability survive a marriage? *Tomorrow* 11 (2): 40-42.
Hasselback, D. 2002. Green code sought. *Financial Post* (Canada), 16 May, FP3.
Hendrik, G., C. Volk, and M. Gilles. 2002. *From economics to sustainomics.* London: WestLB Panmure.
Hillier, B. 2002. Greening the industry: At Dofasco, John Mayberry has shown that what's good for the environment is good for steel. *In Touch: Richard Ivey School of Business* (University of Western Ontario) (Spring): 23-24.
Houlder, V. 2002. Green roads lead to Johannesburg. *Financial Times Survey,* 23 August, 1.
Kennan, G. 1995. On American principles. *Foreign Affairs* 74: 116-26.
Kiernan, M., and J. Martin. 1998. Wake-up call for fiduciaries: Eco-efficiency drives shareholder value. *Today's Corporate Investor* (December): 17-18.
Kommunalkredit Dexia Asset Management. 2004. *Sustainability pays off: An analysis about the stock exchange performance of members of the World Business Council for Sustainable Development (WBCSD).* Vienna: KD Asset Management.
Merrifield, R. 2001. Clean up your acts, FTSE urges. *National Post,* 28 February, C14.
Orlitzky, M. 2004. Meta-analysis reveals clear linkage between CSP and CFP. *Business and the Environment* 15 (12): 2-3.
Papmehl, A. 2002. Investing with a conscience. *Forum* 32: 30-33.
Prial, D. 1999. Social funds a novelty no more. http://cbs.marketwatch.com.
Reed, D. 1998. *Green shareholder value, hype or hit? Sustainable enterprise perspectives.* Washington, DC: World Resources Institute.
Riggs, J. 1998. *Uncovering value: Integrating environmental and financial performance.* Washington, DC: The Aspen Institute.
Russo, M., and P. Fouts. 1997. A resource-based perspective on corporate environmental performance and profitability. *Academy of Management Journal* 40 (3): 1-24.
Savitz, A., M. Besly, and K. Booth. 2002. *2002 sustainability survey report.* Toronto: PricewaterhouseCoopers. Online: http://www.pwcglobal.com/eas.
Schaltegger, S., and F. Figge. 1997. Environmental shareholder value. *WWZ-Study* no. 54. Basel: University of Basel and Bank Sarasin.
Social Investment Forum (SIF). 1999. Socially responsible investing in US tops two trillion dollar mark. http://www.socialinvest.org.
–. 2000. 1999 report on socially responsible investing trends in the United States. http://www.socialinvest.org.

Society of Management Accountants of Canada (SMAC). 1999. *Writing and evaluating sustainable development and environmental reports. Management accounting guideline*. Hamilton, ON: SMAC.

Soyka, P., and S. Feldman. 1998. Investor attitudes toward the value of corporate environmentalism: New survey findings. *Environmental Quality Management* (Autumn): 1-10.

Strong, G.D. 1998. Ethics, environment and enterprise. *Market Perspective: TSE Review* (June): 2-3.

Williams, D., and B. George. 2001. Greening your pension fund. *Benefits Canada* 25 (12): 52-55.

World Commission on Environment and Development (WCED). 1987. *Our common future*. New York: Oxford University Press.

10
Engaging Senior Management on Sustainability

Kevin Brady

The objective of this chapter is to provide some insight on how managers in environmental health and safety (EHS) or sustainable development (SD) functions can more effectively engage senior management on moving their companies to more sustainable forms of production. These insights are based primarily on the consulting work I have done over the past seven years with large multinational companies that were at various stages of integrating environmental and social factors into their business. Another key resource is a series of benchmarking efforts that my company, Five Winds International, has conducted over the last five years.[1] Typically, these benchmarking efforts have focused on understanding the drivers, business case, implementation approaches and challenges, and organizational success factors associated with moving a sustainability initiative forward in an organization.

In my experience, EHS or SD "champions" are typically trying to convince senior managers that the integration of environmental and social considerations into traditional business decision-making processes will result in business benefits (such as improvements in operational efficiency, better management of risks, faster return on investment for capital projects, product innovations, enhanced brand image, etc.). Unfortunately the business climate in which these champions operate (short-term planning horizons, a focus on traditional cost controls, little consideration of non-financial elements in management systems and performance evaluation schemes) is often at odds with effective integration of sustainability, which requires longer planning horizons, more integrated management of financial and non-financial risks and opportunities, and a broader interpretation of the value proposition of the company. In addition, many SD "champions" come from an environmental science background and are not well versed in business language and tools. A further complication is that sustainability is inherently multidisciplinary (requiring a grasp of natural science, business, and social issues) and cross-functional (often requiring an integrated consideration of supply chain management, operations, sales and marketing, and strategy) in nature. Carey Frey makes a similar point in Chapter 6 about the multidisciplinary requirements of the design function.

In spite of these challenges, engaging senior managers is critical to ensuring that society moves toward more sustainable forms of production and consumption. These individuals are the key decision makers in companies and the choices they make can directly or indirectly affect:

- the sustainability of natural resource extraction
- the design of capital projects
- the impacts of manufacturing processes
- flows of labour and capital
- the level of investment (e.g., research and development budgets) for more sustainable materials, products, and services
- energy supply options
- standards of practice related to corporate governance, environmental, and social performance.

Consequently, moving senior managers "up the learning curve" so that they can make more informed choices is a key step in the advancement of a sustainable society.

The Sustainability Journey

The pursuit of sustainable development has long been recognized as a transformational process – a journey rather than a destination (Rowledge et al. 1999). For the purpose of this chapter, an organization can be considered to be on the sustainability journey if it is pursuing the elements indicated in Figure 10.1.[2] At the core are corporate governance procedures that set the business and ethical tone of the company and ensure the proper consideration and management of economic, social, and environmental performance. This model, or variations of it, is at the heart of most corporate interpretations of sustainability, although other terms are often used to describe an organization's activity. These terms include "corporate social responsibility," "corporate citizenship," "sustainable business strategy," "sustainable growth," and so on.

Regardless of the term used, for many organizations the journey typically begins with recognition of the business implications of non-sustainable behaviour. Examples include: the need to manage and eliminate chlorofluorocarbons (CFCs) in the electronics and refrigeration equipment sectors because of the harmful impact of these chemicals on the ozone layer; the recognition by energy companies such as Suncor, BP, and Shell that climate change is a threat to their business that must be responded to; restrictions in the 1980s on the access of North American forest products to the European market based on forest management practices; and the dilution of

Figure 10.1

The sustainable enterprise

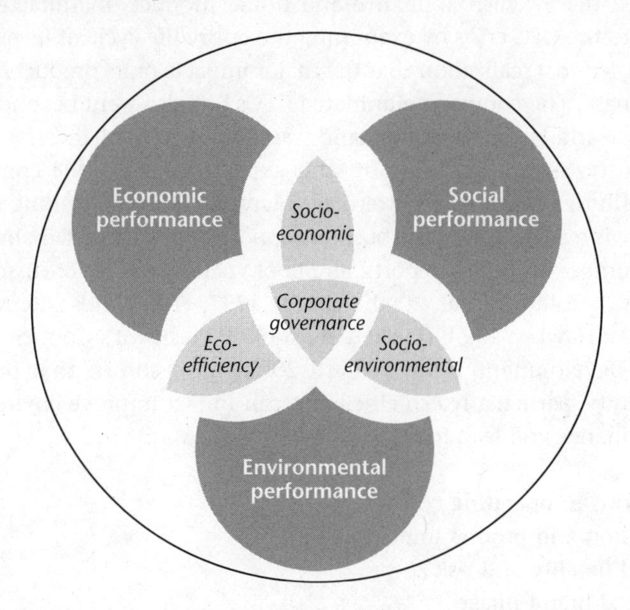

Source: © 2002, Five Winds International.

Nike's brand image due to child labour practices of some of their suppliers in the mid to late 1990s.

In some cases, the response to the "crisis" created by non-sustainable behaviour has been a re-examination of the organization's strategies, policies, and, in some cases, the business model and value proposition of the company. For example, a mining company might shift from viewing itself as an extractive industry to seeing itself as a resource management company that seeks opportunities to recover metals from a diversity of sources, including existing products. This strategy helps them better understand their end-use markets and meet customer needs, while recovering economic and environmental "value" that would have been lost if the metals in the products went to landfill. One interesting example is the California-based partnership between Canadian mining giant Noranda and Hewlett-Packard, in which Noranda "mines" valuable minerals from discarded computers. "In fact, if its facility in Roseville, California, were classified as a Noranda mine, it would account for 5 percent of the copper, 5 percent of the gold, 14 percent of the silver, 37 percent of the platinum group metals mined in Canada" (Nikiforuk 2000, 74). Noranda has opened up a similar metals recycling plant for information and communication technology (ICT) waste in Brampton, Ontario.

In a number of circumstances, this re-examination has started the organization on a path to more sustainable business practices. For example, Electrolux, the Swedish appliance and home products manufacturer, responded to the CFC crisis by examining the entire life cycle of its products. This work led to a realization that the major impacts of its products were in the use phase. The company eliminated CFCs but also went beyond this to become a world leader in energy- and water-efficient appliances.

As industry's understanding of, and experience with, the concept of sustainability grows, a new pattern is emerging. In some leading companies, there is recognition of the business opportunity that sustainability can offer. A number of studies, reports, and books (Anderson 1998; Fusseler and James 1997; Holliday et al. 2000; WBCSD 1997; Five Winds International 2001, 2003; Hawken 1993; Hawken et al. 1999; President's Council for Sustainable Development 1999; Willard 2002) have shown that pursuing sustainability, particularly eco-efficiency, can in fact improve environmental performance and lead to business benefits such as:

- reductions in operating costs
- production and process improvements
- reduced liability and risk
- enhanced brand image
- increased employee morale
- increased opportunities for innovation
- increased opportunity for revenue generation, including new market openings and price premiums
- better supplier management
- better relationships with customers, communities, and regulators.

Further, social and ethical performance aspects of sustainability have increasingly come into focus, and companies are beginning to recognize that investments in "social capital" can also have business benefits (see Chapter 5). These investments often go beyond health and safety and philanthropic activities, and can include stakeholder engagement processes, ethics policies, transparent governance structures and procedures, consistent and fair labour practices, and codes of conduct that apply across the company and in some cases extend throughout the supply chain. These practices can also affect shareholder value by helping a company maintain market access, improve access to resources and capital, protect brand or company reputation, and improve relationships with stakeholders.

Recognition of these potential benefits has led some proactive companies to actively integrate sustainability into their strategic frameworks. These trailblazers have been able to learn from and adapt to new circumstances,

Figure 10.2

Examples of concepts, guidelines, and tools for promoting sustainability in industry

Certification of Environmental Standards (CERES) principles
Cleaner production guides
Corporate social responsibility guidelines (e.g., Canadian Business for Social Responsibility)
Corporate sustainability reporting guidelines (e.g., Global Reporting Initiative)
Design for Environment
Design for Disassembly
Eco-compass
Eco-auditing
Eco-efficiency
Eco-industrial parks
Eco-profiling
Environmental auditing
Environmental management systems
Environmental performance evaluation
Environmental performance indicators
Environmental supply chain management
Global compact
ISO 14000 standards and various national environmental standards
Life cycle assessment, costing, engineering, management, value assessment
Materials stewardship
Natural Step system conditions
Pollution prevention
Product stewardship
Responsible care
Social auditing standards (e.g., SA 8000, AA1000)
Stakeholder engagement processes (e.g., Stakeholder 360)
Sullivan principles

and they have developed new strategies, management systems, programs, and tools that enable them to better integrate environmental, social, and economic factors into their core business functions. This activity is being aided and accelerated by the development, within the research and consulting community, of a growing "toolbox" of management approaches, software tools, and standards (Figure 10.2).

For each successful organization, however, there are more that have had false starts or have failed in their attempt to integrate sustainability considerations into their organizations. A number of attributes contribute to successful integration of SD into companies, including the following (modified from Rowledge et al. 1999):

- a deep understanding within senior management of how sustainable development relates to the company and its entire value chain, its social licence to operate, and ultimately its long-term shareholder value.
- governance structures that ensure full consideration of business risks and opportunities associated with sustainability.
- management systems that ensure that procedures are in place to systematically manage economic, social, and environmental aspects of the company.

- stakeholder dialogue processes that ensure the company not only understands what is expected of them but also benefits from the knowledge and insights that stakeholders can contribute.
- operations and product development processes that ensure optimum productivity of resources and consideration of environmental and social factors up front in decision-making processes (e.g., in capital project development, and in process and product design).
- open, transparent and credible reporting on triple-bottom-line performance – environmental, economic, and social.
- innovative supply management processes where the environmental and economic performance of suppliers is seen as an important component of the organization's performance.
- a toolbox that enables the company to see beyond its gates and understand and assess risks and opportunities along its entire value chain and over longer time horizons. Examples of such tools include scenario analysis, life cycle assessment, design for the environment, life cycle management, and supplier management programs.
- a range of tailor-made communications messages that carry the organization's commitment and performance to key stakeholders, including its customers, its shareholders, and financial markets.

These attributes can enable organizations to identify and manage the business opportunities and realize the business benefits identified above, but they provide only a partial framework for implementing sustainable development within organizations. This is because they do not address the greater challenge, which is how to embed sustainability into the organizational culture and ensure that real change takes place.

According to change management experts Gareth Morgan and Azof Zohar, it is important to understand that large-scale change (such as aligning an organization to pursue sustainability) is not the product of large-scale change programs or of pat formulas. They contend that setting a new direction often unfolds "as the result of crucial, small-scale initiatives that succeed in creating new contexts that break the status quo and allow new streams of new thinking and innovation to occur" (Morgan and Zohar 1995). They believe that a key challenge for managers is to identify "high-leverage" decisions and actions that can push an enterprise in a new direction and that "reverberate and cumulate in their effects." To facilitate the process, managers "must master the art of high-leverage change: by learning how to be driven by a sense of vision while finding 'do-able,' high-impact initiatives that challenge and transform the *status quo;* by learning how to allow one change to build on another and achieve a compounding effect; and by building on opportunities created by random changes that can create unanticipated breakthroughs" (Morgan and Zohar 1995).

A key component of this change effort is engaging senior management and seeking opportunities to align the sustainability efforts with the core business processes and strategy of the company. The need to effectively engage senior management is consistent with the message of the sustainability case study literature, and is one that has been echoed by many companies we have worked with over the past seven years.

Although it is well understood that the support of senior management is critical for moving any major project, activity, or change management initiative forward, what is less understood are the processes for effectively engaging senior management on sustainability. Sustainability differs from many other business issues in that it forces the expansion of traditional decision-making processes to include environmental, economic, and social initiatives in a more integrated or holistic perspective. It also extends the decision-making timeline well beyond traditional business planning cycles. As noted earlier, by definition sustainability is also a multidisciplinary subject that requires insights and inputs from a range of functions within an organization. In the absence of the kind of crisis described earlier, these attributes make engaging management on sustainability a significant challenge.

Some Key Considerations for Engaging Senior Management

Embed Sustainability in the Values, History, and Culture of the Organization

If you examine the mission, vision, or values of most large multinational companies, you will often find references to environmental and social performance along with financial excellence. Many companies have long contributed to the communities in which they operate and have tried to move in the right direction with respect to environmental performance. This is particularly true in companies with community-minded family ownership (e.g., Henkel, Husky Injection Molding, S.C. Johnson). Figure 10.3 provides an example of a declaration of values from S.C. Johnson that reflects some

Figure 10.3

S.C. Johnson's statement of beliefs: "This We Believe"

Employees: We believe our fundamental strength lies in our people.
Consumers: We believe in earning the enduring goodwill of the people who use and sell our products and services.
General Public: We believe in being a responsible leader in the free market economy.
Neighbors and hosts: We believe in contributing to the well being of the countries and communities where we conduct business.
World community: We believe in improving international understanding.

Source: http://www.scjohnson.com (used with permission).

of the key elements of sustainable development. These statements were formalized in 1976, but they can be traced back to 1927 and the values and vision of H.F. Johnson, the son of the company founder.

Sustainability can be very effectively integrated into business strategy when it is linked to the core vision and values of the organization. At Alcoa, the world's leading producer of primary aluminum, fabricated aluminum, and alumina, the corporate vision is to be "the best company in the world." To support this vision, the company has a set of values that reflect environmental, social, and financial excellence. It has developed a model or framework that illustrates how sustainability links to its values, the Alcoa Business System (how it operates), and external stakeholder perspectives. This model positions sustainability as the interface between Alcoa's internal values and society, and it graphically demonstrates to senior management how the company's sustainability efforts can help the company live its values (see Figure 10.4).

Terminology: Sustainable Development, Sustainable Business Strategy, Sustainable Growth, Corporate Social Responsibility?

In the years since the publication of *Our Common Future* (WCED 1987), there has been an explosion of terms, concepts, and tools related to sustainable development. The terminology concerning sustainability is confusing

Figure 10.4

Alcoa sustainability model

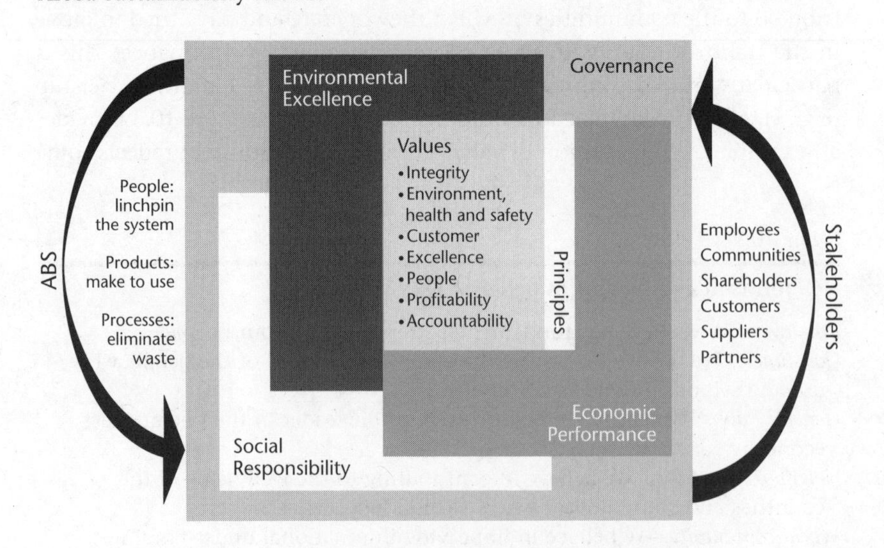

Source: Anita Roper, Director, Sustainability Alcoa (used with permission).

and can be a barrier to effective engagement of senior decision makers, who are often not familiar with the evolution of sustainable development. The choice of terms used should ideally reflect the corporate culture, and concepts must be presented using terms that can be easily explained internally and externally. Large multinational companies operating in developing countries (such as mining and oil and gas companies) are often more comfortable with the term "sustainable development" because development is very much what they do. They develop a resource, create or expand communities around the resource, and often put in place or support basic infrastructure requirements such as roads, health services, and education.

In other cases, companies are less comfortable with SD because they see this as encompassing responsibilities that belong to government and not industry. In such cases, terms like "sustainable growth," "sustainable business strategy," or simply "sustainability" are often used. A growth-oriented manager focusing on financial and production targets may feel more comfortable in a conversation about a sustainable business strategy than in one about corporate social responsibility.

The term "corporate social responsibility" (CSR) has been dominant recently. CSR is essentially a concept whereby companies decide voluntarily to contribute to a better society and a cleaner environment. It includes social responsibility, environmental performance, and, in some interpretations, governance and ethics. In general, when it comes to terminology, it is important to keep in mind the advice of John Kotter, an expert on business leadership. In his view, a key element in the effective communication of a vision is to minimize the use of jargon and keep it simple (Kotter 1996).

Effectively Document and Present Stakeholder Perspectives

Stakeholders' perspectives and expectations regarding company performance are potential levers for engaging senior managers. At one end of the spectrum are customers who can have an immediate impact and can immediately engage management. At the other end are non-governmental organizations that can have an indirect but nevertheless significant impact on a business (for example, through boycott campaigns). More recently, the financial sector is showing interest in sustainability, and inquiries from rating agencies or pension fund managers on sustainability performance can be used to engage management. An important aspect of this activity is the growing linkage between sustainability performance and financial performance. There is a small but growing number of financial sector organizations that are utilizing sustainability to evaluate and direct investment decisions. This type of information can be very effective for engaging the chief financial officer and other senior managers (see Chapter 9).

Stakeholder expectations on sustainability performance are also useful, particularly when the company's performance against these expectations can be shown in comparison with competitors. A caveat concerning presentation of this type of information: it is important to understand the range of expectations, how they relate to the company's value chain, and which ones are of priority because they are coming from key stakeholders (such as regulatory authorities or local communities that can affect the company's licence to operate) or because they relate directly to the company's core business objectives.

Understand Where Management Is on the Sustainability Learning Curve

Professionals in environmental management or sustainability development have typically gone up a long learning curve. This often involves moving from a compliance mindset to pollution prevention, to thinking about industrial systems as complex webs of materials, energy flows, and environmental releases. At the top of this learning curve, these professionals begin to see how the integrated consideration of environmental, social, and economic issues in decision-making processes relates to a variety of core business issues from operational efficiency to innovation, to brand reputation and the overall value proposition of the company.

It is important to take the time to learn where management is on the learning curve in order to devise strategies to move them along. Managers can often have preconceived notions on key issues (such as climate change) that they pick up from the popular media. This can be a barrier to effective engagement and must be overcome through processes that educate them on key issues. Often this is best done in one-on-one sessions where a manager is able to get exposed to new ideas in a non-threatening situation. Many other tactics can be used to bring managers along the learning curve:

- *Subscriptions to a key journal that examines sustainability from a business perspective.* For example, one manager in a European transportation and logistics company subscribed his entire senior management team and board to a business and sustainability magazine in order to expose them to concepts he would be introducing to them.
- *Executive-level training programs on sustainability.* A number of business schools now offer this type of training, the most prominent Canadian example being the Sustainable Enterprise Academy operated by the Schulich School of Business at York University in Toronto. These programs offer senior managers an opportunity to quickly get up to speed on sustainability issues and to interact and exchange views with peers from other companies.
- *Articles by business leaders.* A number of articles and books on sustainability have been written by senior executives. These can be used to engage senior managers, particularly if the article is from an executive in their own

sector. In 2001, the World Business Council for Sustainable Development released a publication on the business case for sustainable development that was signed off by the CEOs of a number of major corporations, including Rio Tinto, DuPont, Shell, BP, Toyota, ST Microelectronics, and Aventis (WBCSD 2001).

- *Personal interactions with industry leaders.* Conversations with, or presentations by, industry peers can also be effective. A number of leading companies that are pursuing sustainability are quite willing to share their stories. This willingness to share can be partly altruistic, in that they believe that the path they have chosen is good for business and society, and partly self-interested, as they recognize that once they have committed to sustainability, getting the word out on what they are doing is important for their stock price, reputation, and company image. Ray Anderson, the CEO of Interface Flooring, is perhaps the most prominent example of industrial leadership.

Identify Business Risk and Opportunity

An important aspect of engaging senior managers is identifying and effectively communicating business risks and opportunities. Of course, these risks and opportunities need to be well documented and presented in both qualitative and quantitative terms. Effective communication can be difficult, as one official described it; "on an emotional level the business case is intuitive but you might not be able to easily present it in terms an accountant can understand." Typically, the business case for sustainability touches on a range of opportunities, some easily quantified (e.g., efficiency improvements) and many not so easily quantified (e.g., the business value of investments in social capital, future value of greenhouse gas (GHG) emissions reductions, technology investments). Senior management often prefers to talk about opportunities rather than problems. One tactic is to show how sustainability efforts can protect value (maintain markets, ensure operating permits, reduce risks), and then show how it can create value (e.g., for a manufacturing company, process and product innovations; or for a resource company, being the developer of choice). Communicating how pursuing sustainability can help protect or create value is often supported by powerful stories that show where companies have failed (e.g., Enron's corporate governance meltdown) or succeeded (Suncor's sustainability-driven corporate turnaround).

Where a company is in the product life cycle (e.g., material supplier, manufacturer, retailer) can also have a significant effect on the risk and opportunity analysis. For example, for upstream companies (e.g., mining, oil and gas), the primary opportunities are related to better capital investments and access to land or resources. For downstream companies (e.g., consumer products and services), opportunities centre on brand image and market access

Figure 10.5

Business risks and opportunities of industrial clients

	Industry Sectors		
	Upstream (e.g., oil and gas, mining)	Midstream (e.g., manufacturing and components)	Downstream (e.g., consumer products, services, and buildings)
Access to resources (land, raw materials, water, etc.)	✓✓✓	✓	✓
Access to capital	✓✓✓	✓	
Access to markets (meeting customer requirements)	✓✓	✓✓	✓✓✓
Licence to operate	✓✓✓	✓	✓
Product and process innovation	✓	✓✓✓	✓✓
Reduced cost and improved benefit	✓✓✓	✓✓✓	✓✓✓
Reduced risk and liability	✓✓	✓✓	✓✓
Employee motivation, retention, and recruitment	✓✓	✓✓	✓
Company image and reputation	✓✓✓	✓✓✓	✓✓✓
Stakeholder relations	✓✓✓	✓✓	✓✓✓
Increased competitiveness	✓	✓✓✓	✓✓✓

Source: © 2002, Five Winds International.

and expansion. Figure 10.5 outlines some internal work we have done at Five Winds to characterize the different business risks and opportunities for our industrial client base as well as governments.

Identify Champions

Having a champion in the senior management ranks who understands the business value of sustainability or its components provides leverage for engaging other managers. Ideally, such a manager would come from outside the traditional environmental function. Support from outside the corporate function is critical as business units or operational groups are often geared toward the core performance goals and targets of the company. If the manager of a business unit is a champion, it is likely that he or she will be able to articulate sustainability in a manner that will resonate with other managers. In cases where the CEO is a champion, it is important to have a sponsor for sustainability within the senior management. The sponsor's role, like that of the champion, is to provide guidance, act as a sounding board for ideas, and help the sustainability team gain access to other senior managers and their leadership teams.

At Rio Tinto Borax, for example, the sustainability effort was initially led by a cross-functional team and then transferred to a set of senior SD champions that included the chief financial officer, the vice president of communications, and the head of environmental health and safety. This is an ideal group of champions to engage the rest of the management team on sustainability because it combines business and environmental expertise with the more outward-looking communications and reporting functions.

Another company that we have studied is Suncor, which has been very successful in its integration of sustainability. In this company, the SD champion reports to the chief financial officer. This places sustainability in the mainstream of the company's day-to-day decision making.

Gain the Support of the Board

A more difficult but potentially powerful leverage point can be a company's board of directors. Theoretically, the board is, in part, responsible for looking out for the long-term viability of the company. One can argue that it is the responsibility of the board to consider risks and opportunities related to sustainability and their relationship to shareholder value. Identifying opportunities to engage board members or committees that are responsible for sustainability, or components of it, is important. Support at the board level can help ensure continuity when there are changes in the management ranks. Outside board members can also be a means of introducing perspectives and experiences that result in a broader understanding of business risk and opportunity.

Ultimately, the governance structure of a company is the glue that will ensure the integrated management of social, economic, and environmental performance. The considerable attention being paid to governance structures of companies due to the evaporation of shareholder value in 2001/2002 provides an opportunity to move corporate social responsibility or sustainability more formally into the realm of the board.

Use Benchmarking

As noted above (see "Effectively Document and Present Stakeholder Perspectives"), benchmarking data are useful for engaging senior managers. Management teams often set targets, such as being best in class or within the first quartile of their sector with respect to financial performance. Being able to demonstrate where the company ranks against competitors on the various elements of sustainability performance can stimulate discussion and spark conversations about why the company's performance is behind or

Figure 10.6

An example of benchmarking data: Overall sustainability ranking of oil and gas companies

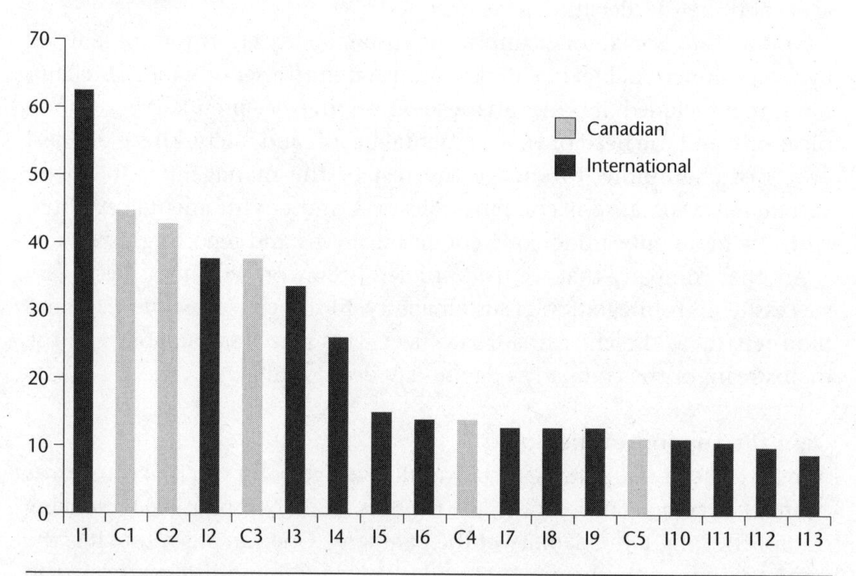

Source: Five Winds International 2003.

ahead of that of competitors. Senior managers tend to be numbers-driven, and being behind on anything does not sit well with them.

Another important set of benchmark data is expectations and trends in a company's customer base. "Radar" functions that gauge customer perspectives and actions on sustainability issues can help demonstrate the business relevance of specific programs (e.g., design, child labour, or GHG management programs) in moving forward. Figure 10.6 shows an example of benchmarking data of interest to senior managers. In this case, oil and gas companies (*x*-axis) were benchmarked against business practices that are implicit in leading international standards and guidance documents on sustainability (e.g., the Global Reporting Initiative's Sustainability Reporting Guidelines). The information enables managers to see where they rank against their peers with respect to stakeholder expectations on performance. The scale on the *y*-axis of the graph indicates that even the leading company meets only about 60 percent of the business practices identified in leading standards and guidance documents.

Speak to Different Functions

Executives responsible for different functions (e.g., marketing, product development, operations, communications, finance, etc.) have issues and priorities that differ in certain ways, even though the company may be moving collectively in one direction. To effectively engage executives from different functions, SD champions must try to understand the specific drivers, business issues, and opportunities for each function. A sales executive can be influenced by a thorough documentation of movement in the customer base to evaluate suppliers' environmental and social performance, while a general manager of a new resource development will be convinced by examples that show how proactive stakeholder engagement can accelerate approvals processes. We recently conducted a project that focused on engaging managers in different business functions, and found the following to be the key steps to effective engagement:

- Understand your audience – What are their priorities, drivers, and business opportunities? What terms do they use?
- Decide what to say – Gather and assemble data in a manner relating to the specific business objectives of the function you are trying to engage.
- Decide how and when to say it – This involves consideration of the form (graphics, stories, detailed business cases) as well as the timing (at a one-on-one meeting; at a leadership team or committee meeting) of the engagement.

Conclusion

Engaging senior decision makers on any issue is a bit of an art, and engaging them on a complex issue such as sustainability is a particular challenge. The ideas presented in this chapter are based on my experience in engaging senior managers and in studying organizational barriers to, and successful factors in, implementing sustainability. One key lesson is that integrating sustainability into an organization's business strategy and culture is a large-scale change management exercise, and therefore has to be done with respect for the organization's past and a deep understanding of the risks and opportunities that sustainability can pose for the company's future. A variety of tactics can be pursued (e.g., linking sustainability to the company's vision or mission statement is one way to tie it directly to the corporate priorities), but engaging the hearts and minds of senior management is the crucial step. Once this has been done effectively, the creative power and managerial expertise of the company leadership can be focused on the pursuit of sustainability.

Notes

1 Five Winds International manages the Product Sustainability Round Table. The Round Table consists of a group of companies that explore and share experiences related to the integration of sustainability and advanced environmental management approaches into business decision-making. For more information, see http://www.fivewinds.com.

2 Being on the sustainability journey should not be confused with actually trying to define a sustainable business. The latter is a much more esoteric and difficult discussion.

References

Anderson, R. 1998. *Mid-course correction: Toward a sustainable enterprise: The Interface model.* Atlanta: Peregrinzilla Press.

Five Winds International. 2001. *The role of eco-efficiency: Global challenges and opportunities in the 21st century.* Ottawa: Five Winds International.

–. 2003. *Corporate social responsibility: Lessons learned – Final summary report.* Prepared for the Interdepartmental Working Group on Corporate Social Responsibility. Ottawa: Five Winds International.

Fusseler, C., and P. James. 1997. *Driving eco-innovation.* London: Pitman.

Hawken, P. 1993. *The ecology of commerce: A declaration of sustainability.* New York: Harper.

Hawken, P., A. Lovins, and H. Lovins. 1999. *Natural capitalism: Creating the next industrial revolution.* Boston: Little, Brown.

Holliday, C. 2001. Sustainable growth, the DuPont way. *Harvard Business Review* (September).

Holliday, C., S. Schmidheiny, and P. Watts. 2002. *Walking the talk: The business case for sustainable development.* Sheffield, UK: Greenleaf.

Kotter, J. 1996. *Leading change.* Boston: Harvard Business School Press.

Morgan, G., and A. Zohar. 1995. Achieving quantum change: Incrementally!!! Schulich School of Business working paper. Toronto: York University.

Nikiforuk, A. 2000. Pure profits. *Canadian Business* (April).

President's Council on Sustainable Development. 1999. *Eco-efficiency Task Force report.* http://clinton4.nara.gov/textonly/PCSD/Publications/TF_Reports/eco-top.html.

Rowledge, R., S. Barton, and K. Brady. 1999. *Mapping the journey: Case studies in developing and implementing sustainable development strategies.* Sheffield, UK: Greenleaf.

Willard, B. 2002. *The sustainability advantage: Seven business case benefits of a triple bottom line.* Gabriola Island, BC: New Society Publishers.

World Business Council for Sustainable Development (WBCSD). 1997. *Environmental performance and shareholder value.* Geneva: WBCSD.

–. 2001. *The business case for sustainable development: Making a difference toward the Johannesburg Summit 2002 and beyond.* Geneva: WBCSD.

World Commission on Environment and Development (WCED). 1987. *Our common future.* New York: Oxford University Press.

Conclusion

11
Whither Sustainable Production? Sustainable Enterprise and the Role of Government

David V.J. Bell and Glen Toner

In Chapter 1, we characterized *Our Common Future* (WCED 1987) as the "seed pod" for the big idea of sustainable development, and subsidiary ideas such as sustainable production. Chapters 2 to 10 have explored many interesting developments in sustainability theory, public policy, and industry practice, which hold promise of future advances in design and production. In Chapter 2, Bob Masterson outlined the major streams of thought that have emerged in the eighteen years between the publication of *Our Common Future* and the publication of the warning signs regarding the continuing loss of natural capital in *Living beyond Our Means* (Millennium Ecosystem Assessment 2005). Without question, this period has spawned many new ideas and inspired leading thinkers who have pushed back the "boundaries of the possible." The eco-efficiency/eco-effectiveness debate around sustainable production has expanded our thinking about what is required as a response to the current condition of industrialization and its environmental, social, and economic consequences.

The Business Council on Sustainable Development propounded the idea of eco-efficiency in *Changing Course: A Global Business Perspective on Development and the Environment* (Schmidheiny 1992) in order to take a progressive business position to the Rio Earth Summit. Eco-efficiency responded to those chapters of *Our Common Future* that criticized the conventional, unsustainable systems of production whose origins can be traced back to the industrial revolution. As Elkington (1998) argued, eco-efficiency was the "Trojan horse" that caused environmental considerations to be brought into executive suite and boardroom discussions of major firms in the early to mid-1990s.

Paul Hawken's seminal *Ecology of Commerce: A Declaration of Sustainability* (1993) had a profound impact on some corporate leaders. Consider Ray Anderson's stunning revelation about the book: "I read it, and it changed my life. It hit me right between the eyes. It was an epiphany. I wasn't halfway through it before I had the vision I was looking for ... for my company, and a powerful sense of urgency to do something to begin to correct the mistakes of the first industrial revolution. Hawken's message was a spear in my chest that is still there" (Anderson 1998, 39-40). This epiphany triggered Anderson's mission of making his company, Interface Flooring, the world's first sustainable enterprise.

The dawning of the millennium focused the minds of leading sustainability thinkers, and the period around the turn of the century generated an explosion of key works. Projections that the planet's population would increase by at least 50 percent, to 9 billion or more, in the twenty-first century (with at least a corresponding increase in industrial production) led many to conclude that efficiency gains alone were not enough. Indeed, critics of eco-efficiency argued that the system of production spawned by the industrial revolution was itself the problem. To give a historical sense of the magnitude of change required in order to move toward sustainability, many authors began to speak in terms of "The Next Industrial Revolution," the title of William McDonough and Michael Braungart's famous *Atlantic Monthly* article in October 1998. In this article, the authors articulated their now-famous critique of eco-efficiency: "Eco-efficiency is an outwardly admirable and certainly well-intended concept, but unfortunately, it is not a strategy for success over the long term, because it does not reach deep enough. It works within the same system that caused the problems in the first place, slowing it down with moral proscriptions and punitive demands. It presents little more than an illusion of change. Relying on eco-efficiency to save the environment will in fact achieve the opposite – it will let industry finish off everything quietly, persistently and completely" (McDonough and Braungart 1998, 85).

They acknowledged that eco-efficiency was a "valuable and laudable tool, and a prelude to what should come next," namely, eco-effectiveness (McDonough and Braungart 1998, 92). They develop this point in *Cradle to Cradle: Remaking the Way We Make Things* (2002) by arguing that the industrial revolution has fundamental design flaws inherent in the linear cradle-to-grave production system. "Doing more with less" still operated within the old flawed model. The design model of the new industrial revolution would be built around cyclical systems of production in which biological nutrients would be designed to decompose and restore natural systems. Non-biological or technical nutrients of the industrial system would be designed to be captured, disassembled, and reused in closed-loop technical cycles, thus reducing the demand for virgin materials from the earth's lithosphere. System design and redesign is at the core of eco-effectiveness. These design principles parallel the four "system conditions" of Karl-Henrik Robèrt's "Natural Step" (2002).

Writing along similar lines, Hawken and Amory and Hunter Lovins entitled their 1999 manifesto *Natural Capitalism: Creating the Next Industrial*

Revolution (1999). This book further developed the argument that the Lovinses had generated with Ernst von Weizsäcker in *Factor Four: Doubling Wealth, Halving Resource Use* (von Weizsäcker et al. 1998), namely, that resource productivity improvements on the scale of Factor 4 or even Factor 10 were required and achievable. *Natural Capitalism* outlines four central strategies: radical resource productivity; biomimicry; service and flow economy; and investing in natural capital (for more on biomimicry, see Benyus 1997).

Many of the tools developed for innovative approaches, such as the Natural Step, triple-bottom-line accounting, industrial ecology, biomimicry, and Design for Environment, are intended to implement elements of this new industrial revolution. In the aftermath of the meltdown of corporate share values and the crisis of corporate ethics flowing from the scandals of the early twenty-first century, focus has shifted to the social dimension of business activity. Corporate social responsibility (CSR), corporate sustainability reporting, and changes in the valuation of firms in the capital markets have been part of the response. Here again, the WBCSD played a leading role in developing business support for CSR, which it defines as "the commitment of business to contribute to sustainable economic development, working with employees, their families, the local community and society at large to improve their quality of life" (WBCSD 2001, 6). All of these developments are signs of the paradigm shift from what Hawken and colleagues (1999) call advanced capitalism to sustainable enterprise.

The contributors to this volume probe many dimensions of this shift. They highlight the environmental harm that results from the conventional production system and cite Einstein's dictum that problems cannot be solved at the level of thinking that created them. Indeed, the issues of system and product design are fundamental. As Carey Frey notes in Chapter 6, "the critical role of design is a theme that continually re-emerges in the sustainable production literature." Design professionals are situated at the interface between users and products. As the linchpin of the economy, they are ultimately responsible for the impacts of their products on the environment. Eighty percent of the economic costs of a product as well as of the environmental and social impacts of the product throughout its life cycle are built in at the design phase. Ecodesign, sustainable product design, and systems thinking are fundamental responses to the limitations of conventional design criteria and production systems.

Sustainability-based education of architects, engineers, and other design professionals will be a key element in moving toward eco-effectiveness. Frey describes university programs that have overcome the limitations of traditional disciplinary assumptions and boundaries to make breakthroughs. The fact that the College of Engineering at a major mainstream university like the University of Michigan has developed the "ConsEnSus"' program

(Concentrations in Environmental Sustainability) to expose engineering students to progressive sustainability-based engineering practices exemplifies this trend. At various paces of change, other schools of engineering, architecture, design, business, and public policy are developing the capacity to provide multidisciplinary, innovative training that over time will improve "the quality of designers' 'mindware' – assets that, unlike physical ones, don't depreciate, but rather ripen with age and experience" (Hawken et al. 1999, 111). The case studies in ConsEnSus show that Ford, DaimlerChrysler, Dow Chemical, Pfizer, BP Amoco, and General Electric are using team-oriented, multidisciplinary approaches to address complex problems in the real world. Business school programs are doing the same.[1]

As "inventors" of the eco-effectiveness idea, McDonough and Braungart (2002) tell the story of the development of their own "mindware" in *Cradle to Cradle*. Despite the fact that the idea of eco-effectiveness is "radical," in that it "goes to the root" in critiquing both the conventional industrial production model and the eco-efficiency alternative, McDonough and Braungart are mainstream players. The fact that they serve as consultants to such industrial revolution icons as Ford shows that the "mainstreaming" of the eco-effectiveness approach is underway. Their role in working with mainstream design professionals to dramatically redesign and redevelop Ford's decrepit Rouge River manufacturing site is quite extraordinary. Frey tells the story of the next stage of their involvement with Ford, working with the eco-effective design team in the creation of the Ford Model U.

The extent to which eco-effectiveness advocates like McDonough and Braungart are working with such dominant industrial players as Nike, Gap, BASF, Visteon, Shaw Industries, Herman Miller, and Ford is significant. Equally important, and indicative of the pace of change, are the number of industry and municipal leaders who are applying the Natural Step principles to their work. The list includes Home Depot, IKEA, Starbucks, Bank of America, Panasonic/Matsushita, Epcor Utilities, the City of Seattle, Cargill Dow, Scandic Hotels, the Whistler Resort, and the Halifax Regional Municipality.

The critical point in any change process occurs when the innovative practices of the sector leaders become the norm of the sector as a whole. Getting to the tipping point where sustainable production practices become the normal practices of the majority of firms in a sector will be a challenge. Nonetheless, as Robert Paehlke argues in Chapter 3, the decoupling of economic output from energy and material throughput (key elements of eco-efficiency) is already happening and indeed is being expedited in most firms as a cost-saving measure as energy and resource commodity costs track upward. Overcoming the conventional industrial system's habits, practices, and processes involves much more, of course. Eco-efficient behaviour – doing more with less – is clearly a step in the right direction. A truly sustainable society, however, will require the broader transition from conventional capitalism to sustainable enterprise (Hart 2005).

Sustainable Enterprise and the Role of Governments

The term "enterprise" (derived from the French *entreprendre,* "to undertake") can be applied to any undertaking or initiative.[2] One dictionary defines enterprise as "an undertaking, especially one of some scope, complication, and risk" (*American Heritage Dictionary*). Given the many environmental and social challenges facing all countries, one could argue that "creating a sustainable society" should be the overarching twenty-first-century enterprise of governments everywhere. The success of this broad, complicated, and risky undertaking will largely determine the future of humankind. Governments at every level and in all regions of the world are beginning to recognize the importance of addressing the challenge of sustainability. Increasingly, the language of sustainable development is being inscribed in policy discourses. Many governments are adopting sustainability-based policies and practices, and some have introduced new policy paradigms, sustainable development institutions, or even constitutional amendments that embed sustainability as a core value for government.

Coming to grips with the challenge of sustainable development is an imperative that goes beyond domestic agendas. No country can be an island of sustainability in a sea of unsustainability. The global context must be factored into the domestic policy agenda. Moreover, it has become increasingly evident that governments acting alone cannot achieve the far-reaching social and economic changes that sustainability will require.

While sustainable development began primarily as a project for governments, as initially set out in *Our Common Future* and later confirmed at the United Nations Conference on Environment and Development (UNCED) in Rio de Janeiro in 1992, the need to engage all sectors of society was always recognized. As has been argued throughout this volume, business obviously must be part of the sustainability solution. Some would even go so far as to agree with Ray Anderson that as the largest institution on earth, business *alone* can lead [toward sustainability] quickly and effectively (Anderson 1998). From this perspective, business must take the lead both domestically and internationally. According to the WBCSD, "the business case for poverty reduction is straightforward. Business cannot succeed in societies that fail" (WBCSD 2001, 14).

Sustainable enterprise entails the production of sustainable goods or services by organizations that are transparent and accountable to all stakeholders (i.e., that operate according to the "process principles" of sustainable development associated with the concept of corporate social responsibility). Sustainable enterprises simultaneously create economic, social, and environmental value while avoiding or minimizing damage to economic, social, or natural capital. To place sustainable enterprise in a broader context, we can use the Chinese concept of "crisis," the word for which consists of two characters, one signifying "danger" and the other "opportunity." Humankind obviously faces multiple *dangers*. The wasteful use of resources creates problems at both ends of the production cycle, inputs and outputs. The natural environment serves as both source and sink. In the 1970s, the work of the Club of Rome focused concern on the source side of the cycle. *The Limits to Growth* (Meadows et al. 1972) used computer simulation models of resource consumption trends plotted against known reserves, and forecast significant resource shortages. Many of the predicted shortages either never appeared because of new source discoveries or price adjustments, or were rendered insignificant because of new technologies that enabled the use of substitute materials. This lulled many into complacency, confident that the dangers identified in *The Limits to Growth* were greatly exaggerated or non-existent.

Indeed, we may not face significant shortages of many materials, although the "end of oil" thesis is gathering some momentum (Roberts 2005). In drawing resources from the lithosphere, we have not gone beyond the first layer of the "onionskin" of the earth. Nonetheless, the way in which we extract resources, the way we use them to manufacture products, what happens when we use these products, and what we do with them when we're finished using them are all producing huge environmental problems. Our manufacturing processes are heavily dependent on fossil fuel, chemicals, and the methodology of "heat, beat, and treat."

The conventional industrial system has serious environmental consequences that include the degradation of many ecosystems. Human actions are contributing to what the World Resources Institute (WRI) has called the "fraying web of life" (WRI 2000). The report of the Millennium Ecosystem Assessment, *Living beyond Our Means: Natural Assets and Human Well-Being* (2005), concluded that "human activity is putting such strain on the natural functions of Earth that the ability of the planet's ecosystem to sustain future generations can no longer be taken for granted ... Nearly two thirds of the services provided by nature to humankind are found to be in decline worldwide. In effect, the benefits reaped from our engineering of the planet have been achieved by running down natural capital assets" (2).

As the research of the 1,360 experts from around the world who did the assessment showed, many industrial activities now have global environmental impacts. Moreover, as the dynamics surrounding the 2005 G8 meeting in Gleneagles, Scotland, underscored, huge deficits exist in the "social dimension," with billions of people living without adequate food, shelter, water, employment, and other necessities of life. We are a long way from fulfilling even the first part of the WCED definition of sustainable development: meeting the needs of the present generation. For some business leaders, this predicament presents an enormous business opportunity – to provide products and services to meet those unmet needs. They see a "fortune at the bottom of the pyramid" (Prahalad and Hart 2002; Hart 2005).

This provides a good segue to the positive face of the crisis of sustainability, the exciting *opportunities* awaiting businesses that can provide solutions to sustainability issues and problems. Indeed, too often advocates of sustainability dwell on the negative, providing an endless litany of problems and challenges. If it is to have any positive effect, however, sustainability must be about solutions – about finding innovative ways of simultaneously improving human well-being and ecosystem health.

In Chapter 5, David Wheeler, Kelly Thomson, and Michael Perkin have provided a series of examples of success stories and a hopeful set of new directions that may lead toward a more sustainable future. Their optimism is tempered by realism. The digital revolution presents enormous potential for creative, exciting outcomes that could contribute to a more equitable and sustainable world, but it might just as easily result in missed opportunities. Success is never guaranteed, as the struggle to deal with electronic waste makes clear.

Wheeler and colleagues, and Keith Newton and John Besley in Chapter 4, explore the multitude of positive and negative impacts of information and communication technologies (ICTs) for sustainable production, and conclude that the current and potential impact is, on balance, positive. Wheeler and colleagues go on to argue that sustainable development is an inclusive societal-level construct that draws on multiple forms of resources, including social and human capital, and emphasizes broad-based value creation in the long term. Indeed, social capital is a critical component in both explaining and achieving sustainable development. They explore various ways in which the inherently strong social capital of the ICT sector could be harnessed to advance both internal and external sustainability. For example, partnerships between hardware and software services firms, civil society organizations and governments could result in both domestic and international initiatives led by Canadian-based ICT firms.

Wheeler and colleagues' conclusion that "the Canadian Government (e.g., Industry Canada and the Canadian International Development Agency) could lead the brokering of partnership arrangements between the Canadian ICT sector and specific geographies in both developed and developing countries, involving civil society organizations and local governments" is very similar to the recommendation made in the *Connecting with the World* report (IDRC 1996), which argued that Canada should make the creation and brokering of knowledge for sustainable development central to Canadian foreign policy and international outreach work (Chapter 4).

Kevin Brady, Blair Feltmate and colleagues, and John Moffet and colleagues explain some of the interesting firm- and sector-level advances we have seen in the late twentieth and early twenty-first centuries. All these authors work closely with firms as they struggle individually and collectively to make breakthroughs in pursuit of sustainable production practices – and, like any transformative process, it is a struggle. As Brady shows in Chapter 10, any number of factors (crisis, a change in leadership, innovative management systems, stakeholder dialogue processes, regulatory pressures, triple-bottom-line considerations) can trigger a firm's pursuit of sustainability. The tremendous growth in the sustainable production toolbox of management approaches, software tools, and standards provides firms with the necessary tools to make the transformation (see Chapter 10). The key to progress is buy-in from the executive suite. Brady provides a strategy for engaging senior management on sustainable production. Different elements of the strategy would apply differently to firms at various points along the sustainability continuum. The interesting point is that we now have enough evidence from firms' experience with the transformative sustainability change process so that analysts like Brady can draw conclusions about what is required to initiate and sustain a high-leverage, high-impact transformation out of the mindset of the old industrial revolution. In *The Sustainability Advantage: Seven Business Case Benefits of a Triple Bottom Line* (2002), Bob Willard builds a compelling business case for why firms ought to commit themselves to sustainability, and outlines the rewards that follow. In *The Next Sustainability Wave: Building Boardroom Buy-In* (2005), he provides a brilliant game plan for addressing the main inhibitors that firms face when contemplating such a change and for overcoming the most likely objections.

Corporate sustainability reporting and environmental performance reporting are two of the new tools of the sustainable production toolbox. In Chapter 7, John Moffet, Stephanie Meyer, and Julie Pezzack show the development and early evolution of corporate sustainability reporting activity in Canada in the early years of the twenty-first century. In the Canadian case, the adoption and Canadianization of the methodology has been a joint public/private sector partnership, an approach that the authors argue is a positive model for overcoming the challenges and limitations of the conventional organizational systems extant in both sectors. After all, government bureaucracies were also designed around the sectoral divisions of the industrial revolution.

Changes to the values, regulations, and norms associated with stock markets and pension funds in several countries have been a key breakthrough for sustainable production. In Chapter 9, Blair Feltmate, Brian Schofield, and Ron Yachnin review the drivers of these changes in the capital markets and some of the key tools that have been developed to assess sustainability challenges and performance. While Brady, Moffet and colleagues, and Feltmate and colleagues provide plenty of evidence that sustainable development is emerging as a value driver in corporate decision making that is positively correlated with share price appreciation, the transformation process within firms and financial markets is neither straightforward nor uncontested. This is entirely normal. Any transformative change process such as that involved with sustainable enterprise will face challenges at many levels, and it is critical that public policy become a supporter of the proactive players and tools.

The Role of Governments

While much of the responsibility for advancing sustainable enterprise might appear to fall on the shoulders of the private sector, Peck and Gibson (2000) make a salient observation about the role of government: "Anticipating rising world demand for sustainable products, services and systems is also an obligation and opportunity for governments. Indeed, there is a crucial role for governments in facilitating the transition to an economy that is much more efficient, much more fair and much less damaging. Governments that lead will be in a stronger position to set the agenda and establish advanced positions for their industries and their citizens. Countries that lag behind will inevitably face increasing competitive disadvantage and lost opportunity."

Support for this proposition is provided by *Green and Gold 2000,* a study by the Institute for Southern Studies (2000). The study finds that American states with the best environmental records also offer the best job opportunities and the most favourable climate for long-term economic development. A similar relationship exists between the worst environmental performers and the lowest-ranked states for economic performance. In short, government has an opportunity and a responsibility to take a leadership role by creating a more suitable habitat in which sustainable enterprise can thrive.

Faced with new pressures in terms of authority, legitimacy, power, and resources, governments have had to rethink their role. For many governments the agenda of the 1990s was dominated by deregulation, downsizing, and deficit reduction. While some have described this process as the "reinvention of government," it might also be described as a process of "creative destruction." One aspect of this change process involves a shift in focus from government to governance. "Government" refers to particular kinds of "public" institutions (the state) that are vested with formal authority to make decisions on behalf of the entire community. "Governance" more broadly refers to the myriad other organizations and institutions, besides government, that make decisions affecting others. Governance encompasses collective decisions made in the public sector, the private sector, and civil society. It suggests the need for collaboration among these sectors to address the kinds of broad, horizontal challenges associated with sustainable development.

Collaboration for sustainability means that increasingly governments must form partnerships – with other levels of government, with the private sector, and/or with civil society organizations. This imperative creates both a danger and an opportunity. The danger is that government will fail to recognize its distinct obligations within such partnerships or will choose this approach even when it is inappropriate.[3] The opportunity for government is to extend the commitment to sustainability through instruments besides the use of law and regulation.

Besides public/private partnerships, several other policy instruments (economic, voluntary, and informational) can complement regulation and legislation. The 2004 *Smart Regulation* report argued that "Smart Regulation offers Canada the opportunity to: support and enable Canadian social, environmental and economic priorities; achieve high standards of protection for citizens; and, support the transition to sustainable development" (External Advisory Committee on Smart Regulation 2004, 6). Choosing the best mix of policy instruments from this wider array of possible options is not an easy task.[4] More research is needed to assess the effectiveness of these various alternatives, and especially to determine the effect they have on sustainable development when used in combination.

Three Key Functions of Government in Promoting Sustainable Enterprise

"Steering" Society toward Sustainability

One of government's key functions is to steer society toward goals that are articulated in public policy. Sometimes governments provide a vision and strategy to help guide policy in the direction of sustainability.[5] For example, the *OECD Environmental Strategy for the First Decade of the 21st Century* (OECD 2001a) is intended to provide clear directions for environmentally sustainable policies in OECD member countries and to guide the future work of the OECD in this area. It was a follow-up to the *OECD Environment Ministers' Shared Goals for Action* (OECD 1998), which invited the OECD to develop a new environmental strategy for the next decade. The OECD environmental performance reviews and the environmental indicators program are used to monitor progress.

Securing the Future, the British government's 2005 Sustainable Development Strategy (United Kingdom 2005), includes five guiding principles (living within environmental limits; ensuring a strong, healthy, and just society; achieving a sustainable economy; promoting good governance; using sound science responsibly) and four priority areas for immediate action (sustainable consumption and production; climate change and energy; natural resource protection and environmental enhancement; sustainable communities) (United Kingdom 2005). David Boyd argues that "Sweden and the Netherlands continue to garner praise. The bold overall goal of both countries is to achieve ecological sustainability within a generation (i.e., by 2020-2025). More importantly, they are implementing the kinds of laws, policies, and actions necessary to attain their ambitious goals. Both have national sustainability strategies that establish measurable objectives and specific timelines for improvement" (2003, 298).

The Canadian government does not have an overall sustainable development strategy but rather has employed an approach in which each federal department is required to produce a sustainable development strategy every three years and report annually on progress toward its implementation. The absence of a government-wide strategy or vision has been a major barrier to change in that it has left individual departmental strategies and their policy and program initiatives disaggregated, uncoordinated, and incoherent (Toner and Frey 2004). Johanne Gelinas, the Commissioner of the Environment and Sustainable Development, encouraged the government in her 2002 report to Parliament to provide a clear picture of what a sustainable Canada would look like twenty years from that time: "If the strategies are to evolve to their full potential, direction and support from the centre of government are essential. The challenge currently faced by each department is like assembling a large jigsaw puzzle without the picture box. Many of the needed pieces are on the table, but it is not clear what picture is meant to emerge" (Commissioner of Environment and Sustainable Development 2002, 5-1).

"Walking the Talk": Practising Sustainability in Government Operations and Purchasing Policies

Public procurement of goods and services covers a range of sectors where environmental issues are important, from the construction and maintenance of highways, public transit systems, and buildings, to the supply of power, water, and sanitation services and the use of vehicles. Excluding the cost of salaries, public procurement expenditures accounted for between 5 percent and 18 percent of GDP in OECD countries in 1997 (OECD 2002). In Canada, the federal government is the largest single purchaser of goods and services, spending some $13 billion each year. By buying goods that, for example, are energy-efficient, produced without using or releasing toxic substances, or easily disassembled for reuse and recycling, governments can significantly reduce the environmental burden of their operations. Equally importantly, green procurement programs can yield indirect benefits by kick-starting markets for innovative goods and services, thereby encouraging businesses and consumers to follow the governments' lead (GVRD 2005).

Many OECD countries have already begun to promote green procurement, for example, through information targeted at procurement officers and the use of environmental pricing and other related financial tools in the evaluation of investments. The OECD urges governments to build on these efforts by providing appropriate policy frameworks and support. Among other things, it recommends establishing appropriate procedures for the identification of greener products; government-wide information; training and technical assistance to facilitate implementation; and the development of indicators to monitor and evaluate programs and policies. OECD environmental performance reviews, undertaken periodically in member countries, will assess the implementation of these steps. The Secretariat of the OECD can also support national initiatives, for example, by coordinating the development of appropriate performance indicators and means of evaluation (OECD 2002).

Creating Appropriate Framework Conditions for Sustainability

One of government's most important functions is to establish the rules that govern social and economic behaviour. If the rules for the market are established properly, governments can help build strong, healthy economies for the twenty-first century that are rooted in sustainable development principles. The converse is also true, as the WBCSD has pointed out. The appropriate framework for a sustainable economy was outlined a decade ago (WBCSD 2001, 4):

> In its 1992 report to the Earth Summit, the then Business Council for Sustainable Development called for a steady, predictable, negotiated move toward full-cost pricing of goods and services; the dismantling of perverse subsidies; greater use of market instruments and less of command-and-control regulations; more tax on things to be discouraged, like waste and pollution and less on things to be encouraged, like jobs (in a fiscally neutral setting); and more reflection of environmental resource use in Standard National Accounts. Other bodies, such as the U.S. President's Council on Sustainable Development, made similar calls. Yet there has been very little political support for such moves from governments, civil society organizations, or frankly, most of business. If basic framework conditions push us all in the wrong directions, then that is the way society will go – until extreme, vociferous forces compel a change.

The BCSD had it right over a decade ago. An integrated policy approach that combines targeted expenditures like green procurement, environmental taxes on emissions and consumption, tax incentives for efficiency upgrades of capital equipment and energy-efficient products, tradable permits, the phasing out of environmentally damaging subsidies, strengthening the effectiveness of voluntary agreements, and regulatory regimes that encourage innovation and positive outcomes is what is required. Many OECD countries are attempting to enhance the integration of this policy toolkit to get the framework conditions "pushing in the right direction."

Ecological fiscal reform (EFR), which reduces taxes on investment and employment and increases charges on pollution and consumption in a revenue-neutral manner, is under discussion in all countries, as it is now generally understood that the tax system is one of the most important elements of the framework for a sustainable economy. The OECD policy brief *Environmentally Related Taxes in OECD Countries: Issues and Strategies* (OECD 2001b) points out that a greater use of market-based instruments is an important framework condition for sustainable development. The brief identifies obstacles to a broader use of such taxes, particularly the fear of loss of sectoral competitiveness, and ways to overcome such problems.

The WBCSD also strongly supports greater reliance on market solutions, pointing out that they are among the most powerful tools available, and that – properly structured – they can be among the least painful. Noting that the market is not good at pricing many environmental assets and services, such as a stable climate or a rich biodiversity and forest cover, the WBCSD also favours such approaches as tradable permit schemes and other efforts to create a market by assigning monetary values to natural resources and natural services (WBCSD 2001, 11).

Canada has been branded a laggard with respect to EFR by many organizations and analysts such as the OECD (2004), the Green Budget Coalition (2005), Boyd (2003), the National Round Table on the Environment and the Economy (2002), and the Canadian Council of Chief Executives (2003). Changes are on the horizon, however. Work on tradable permits is now underway with respect to carbon permits as part of Canada's Kyoto obligations.

Paehlke and a number of other contributors to this volume have argued in support of reforms to Canada's fiscal systems at all levels of government to ensure that "prices tell the truth" by internalizing externalities, and by shifting governmental expenditure and subsidy programs. For example, proponents of suburban sprawl are on the defensive virtually everywhere, and intensification of cities within the boundaries of existing infrastructure is "in" for economic, social, and environmental reasons. An acceleration of the trend of reducing subsidies for sprawl will have major implications for urban form and energy and transportation systems. Urban governments are increasingly reluctant to build infrastructure for new growth unless it is built around public transportation corridors. Significant improvements in the energy efficiency of homes and appliances are the norm. Regulatory responses to air and water quality concerns are having a significant impact on the agriculture, manufacturing, and transportation sectors. Over time, this will result in systemic change and a move up the continuum toward eco-effectiveness as industrial production modifications help transform our societies.

Indeed, the growing acceptance of the end of cheap energy in Canada, combined with the need to reduce emissions of greenhouse gases and smog precursors, will make the energy and transportation fields of particular importance in the early twenty-first century. Fossil fuel prices have tracked upward, with oil reaching $60 a barrel in the summer of 2005. Governments appear increasingly reluctant to continue subsidizing electricity production and wasteful use. The search for renewable energy sources and for alternative energy technologies will have major implications for industrial production, transportation, and urban development systems.[6] The 2000-2005 Canadian federal budgets have increased spending on environmental and sustainable development initiatives and strengthened incentives for alternative energy production, such as wind power, and energy efficiency improvements in capital stock. In response to the criticism at home and abroad that Canada is a laggard in the use of economic instruments to achieve sustainable development objectives, the minister of finance appeared, for the first time, before the House of Commons Standing Committee on Environment and Sustainable Development. Moreover, the 2005 federal budget included a major section outlining the framework the government will use in assessing the growing demand for environmental fiscal policy tools (Canada 2005, 313-27).

In Chapter 8, Mark Jaccard explores examples of increasingly popular policy tools that will help achieve the goals of reducing pollution and greenhouse gas emissions and increasing the supply of renewable energy sources. Cap and trade emission permit systems, renewable energy portfolio standards, and vehicle emission standards are tools that governments can use to send major signals to the marketplace. As Jaccard notes, it is very significant that half of the US states as well as several European countries and Australian states have adopted renewable portfolio standards in their electricity supply systems. Canada has not been a leader in the use of any of these policy tools but there are signs that all three are emerging in Canada in 2005-2006. In and of themselves, they may not be adequate to get Canadians to reduce greenhouse gas emissions, increase the supply of electricity from renewable sources, or purchase more energy-efficient vehicles. However, in combination with increasing energy prices, smart meters or time-of-day pricing for electricity, and tax incentives to reduce the cost of high-efficiency vehicles and to produce wind and other renewable energy sources, they can make a difference.

Conclusion: Where Are We Going?

In *A Short History of Progress* (2004), Ronald Wright eloquently discussed "Gauguin's Three Questions," including "Where are we going?"[7] The contributors to this volume have raised many questions about the challenges that lie just over the horizon. What forces will shape the future of design and production, and what research questions will help us guide the transition to a more sustainable future?

New Issues and Drivers: Sustainability, Security, and the War on Terrorism

No aspect of contemporary society has been untouched by the events of 11 September 2001. What effect has the war on terrorism had on the sustainability agenda?[8] For some policy makers, the former appears to have displaced the latter entirely. The US Environmental Protection Agency's National Advisory Council for Environmental Policy and Technology (NACEPT) acknowledged in its report *The Environmental Future* that the nation's focus has turned "to homeland security and the work of disabling international terrorist organizations" (NACEPT 2002, 1). NACEPT, however, strongly rejected the suggestion that the sustainability challenge can now be ignored, or shifted to the back burner. On the contrary: sustainable development is also essential for reducing social unrest and the danger of international terrorism. No mixture of conditions would be more combustible than rapidly expanding numbers of restless young people living in poverty, without opportunities for improvement, constantly exposed to media images of affluent lifestyles and influenced by new ideologies that preach hatred against America (NACEPT 2002, 7).

Understood properly, sustainable development is not simply a topic to be added to (or deleted from) the agenda of governments. It is the lens through which to view the entire agenda, including the security agenda, in order to develop integrated, coherent strategies for dealing with key issues.[9]

The events of 11 September 2001 had an interesting impact on public opinion. In one post 9-11 US poll, slightly over half the respondents reported that their trust in national government had increased (Putnam 2002). Environics polling in the aftermath of 9-11 indicated that the public in most wealthy countries (including the US) had gone beyond the "crack down on terrorism through armed response" strategy that has dominated the government's reaction to the attacks. The public appears to be amenable to considering the "root causes" of terrorism and addressing poverty and other global social equity issues. If these two findings represent a new trend (rather than simply a blip on the scale), governments' legitimacy will increase but so will public expectations for a more active government role across a whole range of sustainable development policy areas.[10]

Will this translate into support for stronger sustainability commitments and public policies? This is difficult to determine. As discussed in several chapters in this volume, some European countries are moving in this direction, while Canadian and American governments have been very inconsistent in supporting sustainable enterprise and have been slow to implement EFR – or even to remove "perverse subsidies" of unsustainable products and industries. Consequently North American businesses often get mixed signals from governments. In the absence of appropriate framework conditions for sustainability, many companies are adopting a wait-and-see attitude. There may be too few incentives for leaders and too much support for laggards. There is still little evidence of a real shift from traditional public policy paradigms to the new sustainability paradigm.

For public policy to play a larger role in encouraging a transition to sustainable enterprise, policy makers – including both bureaucrats and politicians – will need to show leadership informed by what Tad Homer-Dixon calls "ingenuity." In his book *The Ingenuity Gap* (2000), Homer-Dixon makes the crucially important observation that as social, economic, and political systems become increasingly complex and unpredictable, the impacts of human actions on both the ecosphere and society become difficult to anticipate. We create a plethora of unintended consequences, and we are bedevilled by a rapid increase in "unknown unknowns" – hence the need for a massive increase in ingenuity that "consists not only of ideas for new technologies ... but, more fundamentally of ideas for better institutions and social arrangements" (Homer-Dixon 2000, 2-3).

To increase the supply of ingenuity while tackling the challenges we face will require a paradigm shift, which is precisely what sustainable development aims to accomplish. One aspect of this new paradigm is the practice of "adaptive management," which treats policy as an experimental science/art form that includes continuous (double-loop) learning. Adaptive management is characterized by management that monitors the results of policies and/or management actions and integrates this new learning, adapting policy and management actions as necessary. Policy and management are implemented experimentally, actions are monitored, and the results are integrated to modify policies and management actions, to reassess assumptions in models, and to reassess goals (Holling 1978; Mertsky et al. 2000; Agrawal 2000; Jacobson 2003).

How Can We Move in the Desired Direction?

A Short History of Progress examines the "flight recorders of crashed societies" to determine the causes of societal failure. Wright states (2004, 8): "I want to read these boxes in the hope that we can avoid repeating past mistakes." Drawing on Joseph Tainter's *The Collapse of Complex Societies* (1998), Wright concludes that our civilization shows disturbing features of all three drivers and trends responsible for the tragic outcomes chronicled in Tainter's book: the Runaway Train, the Dinosaur, and the House of Cards. Rapid increases in consumption and waste, population, technology, and the concentration of wealth and power are all examples of the first trend, the Runaway Train. "Hostility to change from vested interests and inertia at all societal levels" (Wright 2004, 129) exemplify the Dinosaur factor. The frightening result of these trends is the tendency of complex civilizations to "fall quite suddenly – the House of Cards effect – because as they reach full demand on their ecologies, they become highly vulnerable to natural fluctuations" (130). Climate change and the cascading effects it could have on ocean currents and weather systems, crop viability, and natural disasters illustrates this possibility. Wright concludes with a note of hope and a dire warning (132):

> We have the tools and the means to share resources, clean up pollution, dispense basic health care and birth control, set economic limits in line with natural ones. If we don't do these things now, while we prosper, we will never be able to do them when times get hard. Our fate will twist out of our hands. And this new century will not grow very old before we enter an age of chaos and collapse that will dwarf all the dark ages in our past. Now is our last chance to get the future right.

What Will It Require to "Get the Future Right" in Relation to Sustainable Production?

An obvious first need is for enhanced dematerialization of production and further development of the new technologies of the second industrial revolution that will decrease rather than augment the environmental impacts of increasing population and affluence. In computing environmental impact, these new technologies must shift to the denominator from the numerator, that is, from $I = P \times A \times T_1$ to $I = (P \times A)/T_2$, where I = impact; P = population; A = affluence; T_1 = technology of the first industrial revolution; T_2 = technology of the next industrial revolution (Anderson 1998, 19).

Second, rapidly increasing urban populations require (re)making cities that work, with all the implications this has for infrastructure, city form, transportation, energy use, and attention to all three aspects of the social dimension of sustainable development (social equity, social capital, and social cohesion).

Third, sustainable production must be linked to more sustainable forms of consumption. This will be a key component of creating a "culture of sustainability" and may entail what Robinson and Tinker (1997) call a "resocialization policy wedge" that will separate the concept of quality of life from high consumption. These imperative requirements for sustainable production suggest the following research questions, which we hope will inform research, including applied research, over the next few years.

- How can we minimize carbon emissions in our economic activities? This challenge entails a complicated web of technical, social, economic, and environmental implications, not all of which are currently understood. We use carbon not only for fuel but also for many consumer goods, including clothing, building materials, and even food production.

- What combination of public policy tools can most expeditiously create the framework conditions to ensure that government actions support sustainable production and enterprise and the transition to a sustainable future?
- What are useful transition strategies to move us toward sustainable production and enterprise? This is both a technical issue and a political/social one. We need to devise better technologies but also address the fears and intransigence of vested interests in both the private and public sectors who benefit or profit from the status quo.
- How can societies effect a cultural shift toward sustainability? There are numerous recent examples of deeply embedded, seemingly fundamental and permanent cultural behaviours and attitudes behaving like a house of cards or a system in transition – flipping from one mode to its opposite in a few brief years. Examples include segregation in the United States, apartheid in South Africa, women's rights in every industrial society, and smoking in many countries in the world. What steps might governments, businesses, and NGOs take to encourage a similar cultural shift toward sustainability?

Notes

1 For more on sustainability education, see the World Resources Institute's BELL (Business-Environment Learning and Leadership) program (WRI 2005), *Beyond Grey Pinstripes 2003: Preparing MBAs for Social and Environmental Stewardship* (Beyond Grey Pinstripes 2005), and Wheeler et al. 2005.

2 Several of the paragraphs in this section have been adapted from Bell 2002.

3 Muldoon and Nadarajah (1999) lament the "decline or retreat of government from its traditional leadership role of defining and promoting economic and social policy goals. Government sees itself as a stakeholder and an equal at the table rather than the 'regulator' protecting the public interest by intervening authoritatively in the 'regulated' industrial sector. The favoured image today is of a partnership with government brokering action where consensus prevails" (54).

4 "The choice of policy instruments is shaped by the characteristics of the instruments, the nature of the problem at hand, governments' past experiences in dealing with the same or similar problems, the subjective preference of the decision-makers, and the likely reaction to the choice by affected social groups" (Howlett and Ramesh 1995, 162). Gunningham and Grabosky (1998) point out that finding the appropriate mix involves much more than merely assembling a "smorgasbord" of options. When different policy instruments are combined, the results can range from "synergy" to "neutralization."

5 One precondition of effective steering is the availability of appropriate feedback information. In relation to governance, this requires the use of indicators and monitoring systems that can measure progress toward sustainability goals and outcomes. Although Chapter 40 of Agenda 21 identified the need for this kind of information for decision making, and despite the emergence since 1992 of what amounts to a growth industry of experts and data on sustainability metrics, integrating these data into decision making at all levels (from the local to the international) remains a challenge. A conference held in 1999 in Costa Rica tackled this challenge directly by looking for ways of integrating sustainability

science and policy. One outcome was the development of the very interesting International Institute for Sustainable Development "dashboard of sustainability" (IISD 2005).

6 The newspaper covering the Canadian Parliament, *The Hill Times,* had a special section on alternative energy in July 2005, in which all the major political parties revealed how they would support alternative and renewable energy systems (11 July 2005, 13-25).

7 The first two questions were "Where do we come from?" and "What are we?"

8 Some believe that security should now be viewed as the "fourth leg" of the sustainability stool. For a discussion, see Bell 2001.

9 The US Army has apparently reached a similar conclusion and is now examining key global "hotspots" through the lens of sustainable development.

10 The Environics (2002) poll of citizens of the G20 countries asked who (governments, companies, or both) was "mainly responsible" for dealing with a number of issues. The percentages citing governments or both were very high across all issues. Examples include "solving social problems" (66 percent governments, 29 percent both); "reducing the gap between rich and poor" (46 percent governments, 46 percent both); and "ensuring industry does not harm the environment" (34 percent governments, 48 percent both.) We are grateful to Chris Coulter of Environics for making the results of this survey available.

References

Agrawal, A. 2000. Adaptive management in trans-boundry protected areas: The Bialowieza National Park and biosphere reserve as a case study. *Environmental Conservation* 27: 326-33.

Anderson, R. 1998. *Mid-course correction: Toward a sustainable enterprise: The Interface model.* Atlanta: Peregrinzilla Press.

Bell, D.V.J. 2001. Sustainability and its implications for peace and security. http://www.sustreport.org/news/comm.html.

–. 2002. The role of government in advancing corporate sustainability. Prepared for the G8 Environmental Futures Forum, Toronto, March.

Benyus, J. 1997. *Biomimicry: Innovation inspired by nature.* New York: William Morrow.

Beyond Grey Pinstripes. 2005. *Beyond Grey Pinstripes 2003: Preparing MBAs for social and environmental stewardship.* http://www.beyondgreypinstipes.org.

Boyd, D. 2003. *Unnatural law: Rethinking Canadian environmental law and policy.* Vancouver: UBC Press.

Canada. 2005. *The budget plan 2005.* Ottawa: Department of Finance. Online: http://www.fin.gc.ca.

Canadian Council of Chief Executives. 2003. *Managing for growth: Fiscal prudence, competitive taxation and smarter spending.* Ottawa: Canadian Council of Chief Executives.

Commissioner of the Environment and Sustainable Development. 2002. *Report to the House of Commons.* Ottawa: Commissioner of the Environment and Sustainable Development.

Elkington, J. 1998. *Cannibals with forks: The triple bottom line of 21st century business.* Gabriola Island, BC: New Society Publishers.

Environics. 2002. *Corporate social responsibility: A global pollster's view.* Toronto: Environics.

External Advisory Committee on Smart Regulation (EACSR). 2004. *Smart regulation: A regulatory strategy for Canada.* Ottawa: EACSR. Online: http://www.pco-pcb.gc.ca/smartreg-regint.

Greater Vancouver Regional District (GVRD). 2005. *Sustainable purchasing guide.* Prepared by Five Winds International. Vancouver: GVRD.

Green Budget Coalition. 2005. *Recommendations for budget 2005.* Ottawa: Green Budget Coalition. Online: http://www.greenbudget.ca.

Gunningham, N., and P. Grabosky. 1998. *Smart regulation: Designing environmental policy.* Oxford: Oxford University Press.

Hart, S. 2005. *Capitalism at the crossroads: The unlimited business opportunities in solving the world's most difficult problems.* Philadelphia: Wharton School Publishing.

Hawken, P. 1993. *The ecology of commerce: A declaration of sustainability.* New York: Harper.

Hawken, P., A. Lovins, and H. Lovins. 1999. *Natural capitalism: Creating the next industrial revolution.* Boston: Little, Brown.

Holling, C.S. 1978. *Adaptive environmental assessment and management.* Chichester, UK: John Wiley and Sons.

Homer-Dixon, T. 2000. *The ingenuity gap.* Toronto: Knopf.

Howlett, M., and M. Ramesh. 1995. *Studying public policy: Policy cycles and policy subsystems.* Toronto: Oxford University Press.

Institute for Southern Studies. 2000. *Green and gold 2000.* http://www.southernstudies.org.

International Development Research Council (IDRC). 1996. *Connecting with the world: Priorities for Canadian internationalism in the 21st century (task force report).* http://www.idrc.ca/en/ev-62072-201-1-DO_TOPIC.html.

International Institute for Sustainable Development (IISD). 2005. *The dashboard of sustainability.* Winnipeg: IISD. Online: http://www.iisd.org/cgsdi/dashboard.asp.

Jacobson, C. 2003. Introduction to adaptive management. http://student.lincoln.ac.nz/am-links/am-intro.html.

McDonough, W., and M. Braungart. 1998. The next industrial revolution. *Atlantic Monthly* (October): 82-92.

–. 2002. *Cradle to cradle: Remaking the way we make things.* New York: North Point Press.

Meadows, D.H., D.L. Meadows, J. Randers, and W. Behrens. 1972. *The limits to growth: A report for the Club of Rome's project on the predicament of mankind.* Washington, DC: Potomac Associates.

Mertsky, V., D. Wegner, and L. Stevens. 2000. Balancing endangered species and ecosystems: A case study of adaptive management in Grand Canyon. *Environmental Management* 25: 579-86.

Millennium Ecosystem Assessment. 2005. *Living beyond our means: Natural assets and human well-being.* http://www.millenniumassessment.org.

Muldoon, P., and R. Nadarajah. 1999. A sober second look. In *Voluntary initiatives: The new politics of corporate greening,* ed. R. Gibson, 51-65. Peterborough, ON: Broadview.

National Advisory Council for Environmental Policy and Technology (US). 2002. *The environmental future: Emerging challenges and opportunities for EPA. A report from the National Advisory Council for Environmental Policy and Technology (NACEPT) September 2002.* http://www.epa.gov/ocempage/nacept/2002_09_epa_nacept_pub_final.pdf.

National Round Table on the Environment and the Economy (NRTEE). 2002. *Toward a Canadian agenda for ecological fiscal reform: First steps.* Ottawa: NRTEE.

Organisation for Economic Co-operation and Development (OECD). 1998. *OECD environment ministers' shared goal for action.* Paris: OECD.

–. 2001a. *OECD environmental strategy for the first decade of the 21st century.* Paris: OECD.

–. 2001b. *Policy brief: Environmentally related taxes in OECD countries: Issues and strategies.* Paris: OECD.

–. 2002. *OECD governments agree to take the lead on buying "green."* Paris: OECD.

–. 2004. *OECD environmental performance reviews: Canada.* Paris: OECD.

Peck, S., and R. Gibson. 2000. Pushing the revolution. *Alternatives* 26 (Winter): 1.

Prahalad, C., and S. Hart. 2002. The fortune at the bottom of the pyramid. *Strategy + Business* 26: 2-14.

Putnam, R. 2002. Bowling together. *The American Prospect* 13: 3. Online: http://www.prospect.org/print/V13/3/putnam-r.html.

Robèrt, K-H. 2002. *The Natural Step story: Seeding a quiet revolution.* Gabriola Island, BC: New Society Publishers.

Roberts, P. 2005. *The end of oil: On the edge of a perilous new world.* Boston: Houghton Mifflin.

Robinson, J., and J. Tinker. 1997. Reconciling ecological, economic and social imperatives: A new conceptual framework. In *Surviving globalism: The social and environmental challenges,* ed. T. Schrecker, 71-94. London: Macmillan.

Schmidheiny, S., with the Business Council for Sustainable Development. 1992. *Changing course: A global business perspective on development and the environment.* Cambridge, MA: MIT Press.

Tainter, J. 1998. *The collapse of complex societies.* Cambridge: Cambridge University Press.

The Hill Times. 2005. Policy briefing: Alternative energy. *The Hill Times,* 11 July, 13-25.

Toner, G., and C. Frey. 2004. Governance for sustainable development: Next stage institutional and policy innovation. In *How Ottawa spends 2004-2005: Mandate change in the Paul Martin era,* ed. G.B. Doern, 198-221. Montreal: McGill-Queen's University Press.

United Kingdom. 2005. *Securing the future: Delivering UK sustainable development strategy.* London: The Stationery Office.

von Weizsäcker, E., A. Lovins, and H. Lovins. 1998. *Factor four: Doubling wealth, halving resource use.* London: Earthscan.

Wheeler, D., A. Zohar, and S. Hart. 2005. Educating senior executives in a novel strategic paradigm: Early experiences of the Sustainable Enterprise Academy. *Business Strategy and the Environment* 14: 172-85.

Willard, B. 2002. *The sustainability advantage: Seven business case benefits of a triple bottom line.* Gabriola Island, BC: New Society Publishers.

–. 2005. *The next sustainability wave: Building boardroom buy-in.* Gabriola Island, BC: New Society Publishers.

World Business Council for Sustainable Development (WBCSD). 2001. *The business case for sustainable development: Making a difference toward the Johannesburg Summit 2002 and beyond.* Geneva: WBCSD.

World Commission on Environment and Development (WCED). 1987. *Our common future.* New York: Oxford University Press.

World Resources Institute (WRI). 2000. People and ecosystems: The fraying web of life. http://www.wri.org.

–. 2005. Business, environment, learning and leadership. http://www.wri.org.

Wright, R. 2004. *A short history of progress.* Toronto: Anansi.

Contributors

David V.J. Bell is Professor Emeritus and former Dean of the Faculty of Environmental Studies, York University.

John C. Besley is an assistant professor of risk communication in the School of Journalism and Mass Communication at the University of South Carolina, Columbia.

Kevin Brady is Founding Partner and Director of Corporate Sustainability Services, Five Winds International, Ottawa.

Blair W. Feltmate is Director of Sustainable Development, Ontario Power Generation.

Carey Frey is a Research Associate in the Carleton Research Unit in Innovation, Science and Environment, Carleton University.

Mark Jaccard is a Professor in the School of Resource and Environmental Management, Simon Fraser University.

Bob Masterson is a Senior Consultant at Stratos Inc., Ottawa.

Stephanie Meyer is a Principal at Stratos Inc., Ottawa.

John Moffet is a Principal at Stratos Inc., Ottawa, and is on Executive Interchange as Director General, Systems and Priorities Directorate, at Environment Canada from July 2005 to July 2006.

Keith Newton is an Adjunct Professor in the School of Public Policy and Administration, Carleton University.

Robert Paehlke is Professor and Past Chair, Environmental and Resource Studies, Trent University.

Michael A. Perkin was an Internet entrepreneur.

Julie Pezzack is a Senior Consultant at Stratos Inc., Ottawa.

Brian A. Schofield is a Portfolio Manager at Scotia Cassels Investment Counsel, Toronto.

Kelly Thomson is an Assistant Professor at the Atkinson Faculty for Liberal and Professional Studies, York University.

Glen Toner is Professor and Graduate Supervisor in the Innovation, Science and Environment Stream, School of Public Policy and Administration, Carleton University.

David Wheeler is Chair and Director of the Erivan K. Haub Program in Business and Sustainability, York University.

Ron Yachnin is a Principal at Yachnin and Associates.

Index

Printed and bound in Canada by Friesens

Set in Stone by Artegraphica Design Co. Ltd.

Copy editor: Francis J. Chow

Indexer: Annette Lorek